Praise for *The Technological Republic*

The instant *Sunday Times* and *New York Times* bestseller

'[*The Technological Republic*] helps explain the sudden and extraordinary change of worldview that has seized much of the US tech elite ... A fascinating, if at times disturbing, insight into the reassertion of US hard power'
Financial Times

'*The Technological Republic* sets out [Karp and Zamiska's] vision ... for how to meet that challenge [of AI]. It is far too important to ignore' *Times Literary Supplement*

'A surprising book ... nuanced and provocative. Even the most ardent pacifists should consider the arguments in it' *Irish Times*

'The AI manifesto inspiring Keir Starmer's government'
The Times

'A bold and ambitious work, *The Technological Republic* reminds us of a time when technological progress answered a national calling. It is essential reading in the age of AI' Eric Schmidt, former CEO of Google

'[Karp] has built a unique business. *The Technological Republic* combines fascinating insights into Palantir's mode of operation [with] Karp's uncompromisingly national-liberal political philosophy ... "Silicon Valley, awake!" is Karp and Zamiska's message' Niall Ferguson, author of *The Ascent of Money*

'Karp's resilience, patriotism and depth of experience in our rapidly changing world provide instructive lessons and intellectual arguments for all of us' Jamie Dimon, CEO of JPMorgan Chase

'Karp's book might be titled *A Freethinker's Manifesto*. He decries the arrogance and small-mindedness of Silicon Valley and explains his passionate commitment to defending the West and its cultural values. Karp is a polymath: he and his co-author, Nicholas Zamiska, take the reader on an intellectual tour from anthropology to art and music to history and philosophy to explain what matters for our survival and success' David Ignatius, author of *Phantom Orbit*

'Not since Allan Bloom's astonishingly successful 1987 book *The Closing of the American Mind* ... has there been a cultural critique as sweeping' *Washington Post*

'Equal parts company lore, jeremiad and homily ... The primary target of *The Technological Republic* is not a nation that has failed Silicon Valley. It is more cogent and original as a story about how Silicon Valley has failed the nation' *New Yorker*

'No less ambitious than a new treatise in political theory ... A cri de coeur that takes aim at the tech industry for abandoning its history of helping America and its allies' *Wall Street Journal*

'This is an extremely important book ... Alexander Karp is a brilliant out-of-consensus visionary who has built one of the most consequential companies in America ... his insight [is] both provocative and invaluable' Stanley Druckenmiller

'In rendering such a deep and subtle meditation about the role of morality in the private sector and the world of power, *The Technological Republic* may be the most exhilarating political treatise of the decade' *Quilette*

'As clear and bracing as reveille ... with engaging storytelling' *Washington Post*

ALEXANDER C. KARP

Alexander C. Karp is co-founder and CEO of Palantir Technologies Inc. The company, established in Palo Alto, California, in 2003, builds software and artificial intelligence capabilities that are used by defence and intelligence agencies in the United States and allied nations around the world, as well as companies across the commercial sector. Karp is a graduate of Haverford College and Stanford Law School. He earned his doctorate in social theory from Goethe University in Frankfurt, Germany.

NICHOLAS W. ZAMISKA

Nicholas W. Zamiska is head of corporate affairs and legal counsel to the office of the CEO at Palantir. He also serves on the board of directors of the Palantir Foundation for Defense Policy & International Affairs. Zamiska received his J. D. from Yale Law School and is a graduate of Yale College. He was born in New York City.

ALEXANDER C. KARP &
NICHOLAS W. ZAMISKA

The Technological Republic

Hard Power, Soft Belief and
the Future of the West

VINTAGE

1 3 5 7 9 10 8 6 4 2

Vintage is part of the Penguin Random House group of companies

Vintage, Penguin Random House UK, One Embassy Gardens,
8 Viaduct Gardens, London SW11 7BW

penguin.co.uk/vintage
global.penguinrandomhouse.com

First published in Vintage in 2026
First published in hardback by The Bodley Head in 2025
First published in the United States of America by Crown Currency in 2025

Copyright © Alexander C. Karp and Nicholas W. Zamiska 2025

Designed by Aubrey Khan

The moral right of the authors has been asserted

Penguin Random House values and supports copyright. Copyright fuels creativity, encourages diverse voices, promotes freedom of expression and supports a vibrant culture. Thank you for purchasing an authorised edition of this book and for respecting intellectual property laws by not reproducing, scanning or distributing any part of it by any means without permission. You are supporting authors and enabling Penguin Random House to continue to publish books for everyone. No part of this book may be used or reproduced in any manner for the purpose of training artificial intelligence technologies or systems. In accordance with Article 4(3) of the DSM Directive 2019/790, Penguin Random House expressly reserves this work from the text and data mining exception.

Printed and bound in Great Britain by Clays Ltd, Elcograf S.p.A.

The authorised representative in the EEA is Penguin Random House Ireland, Morrison Chambers, 32 Nassau Street, Dublin D02 YH68

A CIP catalogue record for this book is available from the British Library

ISBN 9781529945409

Penguin Random House is committed to a sustainable future for our business, our readers and our planet. This book is made from Forest Stewardship Council® certified paper.

To those who seek

to move the hearts of others

and know their own

•

You will never touch the hearts of others,
if it does not emerge from your own.

("Werdet ihr nie Herz zu Herzen schaffen,
Wenn es euch nicht von Herzen geht.")

—JOHANN WOLFGANG VON GOETHE

The power to hurt is bargaining power.
To exploit it is diplomacy—
vicious diplomacy, but diplomacy.
—THOMAS SCHELLING

•

Fundamentalists rush in
where liberals fear to tread.
—MICHAEL SANDEL

Contents

List of Figures ... *xiii*
Preface ... *xv*

Part I
The Software Century ... *1*

One	Lost Valley	3
Two	Sparks of Intelligence	16
Three	The Winner's Fallacy	29
Four	End of the Atomic Age	37

Part II
The Hollowing Out of the American Mind ... *55*

Five	The Abandonment of Belief	57
Six	Technological Agnostics	69
Seven	A Balloon Cut Loose	83
Eight	"Flawed Systems"	97
Nine	Lost in Toyland	103

CONTENTS

Part III
The Engineering Mindset ... 113

Ten	The Eck Swarm	115
Eleven	The Improvisational Startup	122
Twelve	The Disapproval of the Crowd	130
Thirteen	Building a Better Rifle	139
Fourteen	A Cloud or a Clock	156

Part IV
Rebuilding the Technological Republic ... 169

Fifteen	Into the Desert	171
Sixteen	Piety and Its Price	179
Seventeen	The Next Thousand Years	190
Eighteen	An Aesthetic Point of View	205

Acknowledgments	219
Notes	221
Bibliography	261
Art Credits	285
Index	287

List of Figures

Figure 1   The Unicorn Drawing Test

Figure 2   Battle-Related Deaths Per 100,000 People Worldwide (1946 to 2016)

Figure 3   Defense Spending as a Percentage of GDP: United States and Europe (1960 to 2022)

Figure 4   Total Factor Productivity Growth in the United States (1900 to 2014)

Figure 5   Percentage of Harvard Graduates Bound for Finance or Consulting (1971 to 2022)

Figure 6   The Very Long Term: Estimated GDP Per Capita Worldwide (AD 1 to 2003)

Figure 7   The Huntington-Wallace Line

Figure 8   Western Empires: Share of Territory and Global Economic Output

Figure 9   Locations of Potential Nesting Sites as Indicated by Honeybee Dances in the Eck Swarm

Figure 10 The Asch Conformity Experiment

Figure 11 Percentage of Members of U.S. Congress Who Have Served in the Military

Figure 12 Accuracy of Predictions Made by "Foxes" and "Hedgehogs" in Philip Tetlock's Review of 284 Experts

Figure 13 Support for U.S. Major League Baseball Teams as of 2014

Figure 14 *Ulysses and the Sirens* by Herbert James Draper, 1909

Figure 15 The Founder Premium: Total Return of Founder-Led Companies vs. Others (1990 to 2014)

Preface

THIS BOOK IS THE PRODUCT of a nearly decade-long conversation between its authors regarding technology, our national project, and the perilous political and cultural challenges that we collectively face.

A moment of reckoning has arrived for the West. The loss of national ambition and interest in the potential of science and technology, and resulting decline of government innovation across sectors, from medicine to space travel to military software, have created an innovation gap. The state has retreated from the pursuit of the kind of large-scale breakthroughs that gave rise to the atomic bomb and the internet, ceding the challenge of developing the next wave of pathbreaking technologies to the private sector—a remarkable and near-total placement of faith in the market. Silicon Valley, meanwhile, turned inward, focusing its energy on narrow consumer products, rather than projects that speak to and address our greater security and welfare.

The current digital age has been dominated by online advertising and shopping, as well as social media and video-sharing platforms. The grandiose rallying cry of a generation of founders in Silicon Valley was simply to build. Few asked what needed to be built, and why. For decades, we have taken this focus—and indeed obsession in many cases—by the technology industry on consumer culture for granted, hardly questioning the direction, and we think misdirection, of capital and talent to the trivial and ephemeral. Much of what passes for innovation today, of what attracts enormous amounts of talent and funding, will be forgotten before the decade is out.

The market is a powerful engine of destruction, creative and otherwise, but it often fails to deliver what is most needed at the right time. The Silicon Valley giants that dominate the American economy have made the strategic mistake of casting themselves as existing essentially outside the country in which they were built. The founders who created these companies in many cases viewed the United States as a dying empire, whose slow descent could not be allowed to stand in the way of their own rise and the new era's gold rush. Many of them essentially abandoned any serious attempt to advance society, to ensure that human civilization kept inching up the hill. The prevailing ethical framework of the Valley, a techno-utopian view that technology would solve all of humanity's problems, has devolved into a narrow and thin utilitarian approach, one that casts individuals as mere atoms in a system to be managed and contained. The vital yet messy questions of what constitutes a good life, which collective endeavors society should pursue, and what a shared and national identity can make possible have been set aside as the anachronisms of another age.

We can—we must—do better. The central argument that we advance in the pages that follow is that the software industry should rebuild its relationship with government and redirect its effort and attention to constructing the technology and artificial intelligence capabilities that will address the most pressing challenges that we collectively face. The engineering elite of Silicon Valley has an affirmative obligation to participate in the defense of the nation and the articulation of a national project—what is this country, what are our values, and for what do we stand—and, by extension, to preserve the enduring yet fragile geopolitical advantage that the United States and its allies in Europe and elsewhere have retained over their adversaries. It is, of course, the protection of individual rights against state encroachment that took its modern shape within "the West"—a concept that has been discarded by many, almost casually—without which the dizzying ascent of Silicon Valley would never have been possible.

The rise of artificial intelligence, which for the first time in history presents a plausible challenge to our species for creative supremacy in the world, has only heightened the urgency of revisiting questions of national identity and purpose that many had thought could be safely cast aside. We might have muddled through for years if not decades, dodging these more essential matters, if the rise of advanced AI, from large language models to the coming swarms of autonomous robots, had not threatened to upend the global order. The moment, however, to decide who we are and what we aspire to be, as a society and a civilization, is now.

Others might prefer or advocate for a more careful and deliberate division between the domains and concerns of the private and the public sectors. The blending of business and national purpose, of the discipline that the market can provide with an interest in the collective good, makes many uneasy. But purity comes at a cost. We believe that the reluctance of many business leaders to venture into, in any meaningful way and aside from the occasional and theatrical foray, the most consequential social and cultural debates of our time—including those regarding the relationship between the technology sector and the state—should give us pause. The decisions we collectively face are too consequential to be left unchallenged and unexamined. Those involved in constructing the technology that will animate and make possible nearly every aspect of our waking lives have a responsibility to expose and defend their views.

Our broader hope is that this book prompts a discussion of the role Silicon Valley can and should play in the advancement and reinvention of a national project, both in the United States and abroad—of what, beyond a firm and uncontroversial commitment to liberalism and its values, including the advancement of individual rights and fairness, constitutes our shared vision of the community to which we belong.

We recognize that a political treatise of this nature is an unusual project for those in the private sector to undertake. But the stakes are high, and growing. The technology industry's current reluctance to engage with these fundamental questions has deprived us of a positive vision for what this country or any other can and should be in an era of increasing technological change and risk. We also believe that the values of the engineering culture that gave rise to Silicon Valley, including its obsessive focus on outcomes and disinterest in theater and posturing—while complex and imperfect—will in the end prove vital to our ability to advance our national security and welfare.

Too many leaders are reluctant to venture into the discussion, to articulate genuine belief—in an idea, a set of values, or a political project—for fear that they will be punished in the contemporary public sphere. A significant subset of our leaders, elected and otherwise, both teach and are taught that belief itself is the enemy and that a lack of belief in anything, except oneself perhaps, is the most certain path to reward. The result is a culture in which those responsible for making our most consequential decisions—in any number of public domains, including government, industry, and academia—are often unsure of what their own beliefs are, or more fundamentally if they have any firm or authentic beliefs at all.

We hope that this book, including by its very existence, suggests that a far richer discourse, a more meaningful and nuanced inquiry into our beliefs as a society, shared and otherwise, is possible—and, indeed, imperative. Those in the private sector should not cede this terrain to others in academia and elsewhere out of a perceived lack of authority or expertise. Palantir itself is an attempt—imperfect, evolving, and incomplete—at constructing a collective enterprise, the creative output of which blends theory and action. The company's deployment of its software and its work in the world constitute the action. This book attempts to offer the beginnings of an articulation of the theory.

ACK AND NWZ
NOVEMBER 2024

Part I

The Software Century

Chapter One

Lost Valley

SILICON VALLEY HAS LOST ITS WAY.

The initial rise of the American software industry was made possible in the first part of the twentieth century by what would seem today to be a radical and fraught partnership between emerging technology companies and the U.S. government. Silicon Valley's earliest innovations were driven not by technical minds chasing trivial consumer products but by scientists and engineers who aspired to see the most powerful technology of the age deployed to address challenges of industrial and national significance. Their pursuit of breakthroughs was intended not to satisfy the passing needs of the moment but rather to drive forward a much grander project, channeling the collective purpose and ambition of a nation. This early dependence of Silicon Valley on the nation-state and indeed the U.S. military has for the most part been forgotten, written out of the region's history as an inconvenient and dissonant fact—one that clashes with the Valley's conception of itself as indebted only to its capacity to innovate.

In the 1940s, the federal government began supporting an array of research projects that would culminate in the development of novel pharmaceutical compounds, intercontinental rockets, and satellites, as well as the precursors to artificial intelligence. Indeed, Silicon Valley once stood at the center of American military production and national security. Fairchild Camera and Instrument Corporation, whose semiconductor division was founded in Mountain View, California, and made possible the first primitive personal computers, built reconnaissance equipment for spy satellites used by the Central Intelligence Agency beginning in the late 1950s. For a time after World War II, all of the U.S. Navy's ballistic missiles were produced in Santa Clara County, California. Companies such as Lockheed Missile & Space, Westinghouse, Ford Aerospace, and United Technologies had thousands of employees working in Silicon Valley on weapons production through the 1980s and into the 1990s.

This union of science and the state in the middle part of the twentieth century arose in the wake of World War II. In November 1944, as Soviet forces closed in on Germany from the east and Adolf Hitler prepared to abandon his Wolf's Lair, or *Wolfsschanze*, his eastern front headquarters in the north of present-day Poland, President Franklin Roosevelt was in Washington, D.C., already contemplating an American victory and the end of the conflict that had remade the world. Roosevelt sent a letter to Vannevar Bush, the son of a pastor who had become the head of the U.S. Office of Scientific Research and Development. Bush was born in 1890 in Everett, Massachusetts, just north of Boston. Both his father and his grandfather had grown up in Provincetown at the far end of Cape Cod. In the letter, Roosevelt described the "unique experiment" that the United States had undertaken during the war to leverage science in service of military ends. Roosevelt anticipated the next era—and partnership between national government and private industry—with precision. He wrote that there is "no reason why the lessons to be found in this experiment"—that is, directing the resources of an emerging scientific establishment to help wage the most significant and violent war that the world had ever known—"cannot be profitably employed in times of peace." His ambition was clear. Roosevelt intended to see that the machinery of the state—its power and prestige, as well as the financial resources of the newly victorious nation and emerging hegemon—would spur the scientific community forward in service of, among other things, the advancement of public health and national welfare. The challenge was to ensure that the engineers and researchers who had directed their attention to the industry of war—and particularly the physicists, who as Bush noted had "been thrown most violently off stride"—could shift their efforts back to civilian advances in an era of relative peace.

The entanglement of the state and scientific research both before and after the war was itself built on an even longer history of connection between innovation and politics. Many of the earliest leaders of the American republic were themselves engineers, from Thomas Jefferson, who designed sundials and studied writing machines, to Benjamin Franklin, who experimented with and constructed everything from lightning rods to eyeglasses. Franklin was not someone who dabbled in science. He was an engineer, one of the most productive in the century, who happened to become a politician. Dudley Herschbach, a Harvard professor and chemist, has observed that the Founding Father's research into electricity "was recognized as ushering in a scientific revolution comparable to those wrought by Newton in the previous century or by Watson and Crick in ours." For Jefferson, science and natural history were his "passion," he wrote in a letter to a federal judge in Kentucky in 1791, while politics was his "duty." Some fields were so new that nonspecialists could aspire to make plausible contributions to them. James Madison dissected an American weasel and took nearly forty measurements of the animal in order to compare it with European varieties of the species, as part of an investigation into a theory, advanced by the French naturalist Georges-Louis Leclerc in the eighteenth century, that animals in North America had degenerated into smaller and weaker versions of their counterparts across the ocean.

Unlike the legions of lawyers who have come to dominate American politics in the modern era, many early American leaders, even if not practitioners of science themselves, were nonetheless remarkably fluent in matters of engineering and technology.* John Adams, the second president of the United States, by one historian's account was focused on steering the early republic away from "unprofitable science, identifiable in its focus on objects of vain curiosity," and toward more practical forms of inquiry, including "applying science to the promotion of agriculture." The innovators of the eighteenth and nineteenth centuries were often polymaths whose interests diverged wildly from the contemporary expectation that depth, as opposed to breadth, is the most effective means of contributing to a field. The term "scientist" itself was only coined in 1834, to describe Mary Somerville, a Scottish astronomer and mathematician; prior to that, the blending of pursuits across physics and the humanities, for instance, was so commonplace and natural that a more specialized word had not been needed. Many had little regard for the boundary lines between disciplines, ranging from areas of study as seemingly unrelated as linguistics to chemistry, and zoology to physics. The frontiers and edges of science were still in that earliest stage of expansion. As of 1481, the library at the Vatican, the largest in Europe, had around thirty-five hundred books and documents. The limited extent of humanity's collective knowledge made possible and encouraged an interdisciplinary approach that would almost be certain to stall an academic career today. That cross-pollination, as well as the absence of a rigid adherence to the boundaries between disciplines, was vital to a willingness to experiment, and to the confidence of political leaders to opine on engineering and technical questions that implicated matters of government.

---

* We have, in the modern era, crowded out technical minds from electoral office. There are notable exceptions. Margaret Thatcher, for example, worked as a chemist at a plastics firm before becoming the British prime minister, and Angela Merkel earned a doctorate in quantum chemistry in East Germany prior to serving as chancellor. Yet contemporary democratic regimes have not placed scientists at their center. A survey conducted in 2023 found that only 1.3 percent of state legislators in the United States were either scientists or engineers.

The rise of J. Robert Oppenheimer and dozens of his colleagues in the late 1930s only further situated scientists and engineers at the heart of American life and the defense of the democratic experiment. Joseph Licklider, a psychologist whose work at the Massachusetts Institute of Technology anticipated the rise of early forms of AI, was hired in 1962 by the organization that would become the U.S. Defense Advanced Research Projects Agency—an institution whose innovations would include the precursors to the modern internet as well as the global positioning system. His research for his now classic paper "Man-Computer Symbiosis," which was published in March 1960 and sketched a vision of the interplay between computing intelligence and our own, was supported by the U.S. Air Force. There was a closeness, and significant degree of trust, in the relationships between political leaders and the scientists on whom they relied for guidance and direction. Shortly after the launch by the Soviet Union of the satellite Sputnik in October 1957, Hans Bethe, the German-born theoretical physicist and adviser to President Dwight D. Eisenhower, was called to the White House. Within an hour, there was agreement on a path forward to reinvigorate the American space program. "You see that this is done," Eisenhower told an aide. The pace of change and action in that era was swift. NASA was founded the following year.

By the end of World War II, the blending of science and public life—of technical innovation and affairs of state—was essentially complete and unremarkable. Many of these engineers and innovators would labor in obscurity. Others, however, were celebrities in a way that might be difficult to imagine today. In 1942, as war spread across Europe and the Pacific, an article in *Collier's* introduced Vannevar Bush, who would help found the Manhattan Project but was at the time a little-known engineer and government bureaucrat, to the magazine's readership of nearly three million, describing Bush as "the man who may win the war." An interest in those untangling the most fundamental mysteries of the physical world had been growing for decades on both sides of the Atlantic. Marie Curie sent a letter to her brother in 1903, shortly after discovering radium and winning the Nobel Prize, her first of two, noting the onslaught of requests from journalists. "One would like to dig into the ground somewhere to find a little peace," she wrote. Similarly, Albert Einstein was not only one of the twentieth century's greatest scientific minds but also one of its most prominent celebrities—a popular figure whose image and breakthrough discoveries that so thoroughly defied our intuitive understanding of the nature of space and time routinely made front-page news. And it was often the science itself that was the focus of coverage.

This was the American century, and engineers were at the heart of the era's ascendant mythology. The pursuit of public interest through science and engineering was considered a natural extension of the national project, which entailed not only protecting U.S. interests but moving society, and indeed civilization, up the hill. And while the scientific community required funding and extensive support from the government, the modern state was equally reliant on the advances that those investments in science and engineering produced. The technical outperformance of the United States in the twentieth century—that is, the country's ability to reliably deliver economic and scientific advances for the public, from medical breakthroughs to military capabilities—was essential to its credibility. As Jürgen Habermas has suggested, a failure by leaders to deliver on implied or explicit promises to the public has the potential to provoke a crisis of legitimacy for a government. When emerging technologies that give rise to wealth do not advance the broader public interest, trouble often follows. Put differently, the decadence of a culture or civilization, and indeed its ruling class, will be forgiven only if that culture is capable of delivering economic growth and security for the public. In this way, the willingness of the engineering and scientific communities to come to the aid of the nation has been vital not only to the legitimacy of the private sector but to the durability of political institutions across the West.

* * *

The modern incarnation of Silicon Valley has strayed significantly from this tradition of collaboration with the U.S. government, focusing instead on the consumer market, including the online advertising and social media platforms that have come to dominate—and limit—our sense of the potential of technology. A generation of founders cloaked themselves in the rhetoric of lofty and ambitious purpose—indeed their rallying cry to *change the world* has grown lifeless from overuse—but often raised enormous amounts of capital and hired legions of talented engineers merely to build photo-sharing apps and chat interfaces for the modern consumer. A skepticism of government work and national ambition took hold in the Valley. The grand, collectivist experiments of the earlier part of the twentieth century were discarded in favor of a narrow attentiveness to the desires and needs of the individual. The market rewarded shallow engagement with the potential of technology, as startup after startup catered to the whims of late capitalist culture without any interest in constructing the technical infrastructure that would address our most significant challenges as a nation. The age of social media platforms and food delivery apps had arrived. Medical breakthroughs, education reform, and military advances would have to wait.

For decades, the U.S. government was viewed in Silicon Valley as an impediment to innovation and a magnet for controversy—the obstacle to progress, not its logical partner. The technology giants of the current era long avoided government work. The level of internal dysfunction within many state and federal agencies created seemingly insurmountable barriers to entry for outsiders, including the insurgent startups of the new economy. In time, the tech industry grew disinterested in politics and broader communal projects. It viewed the American national project, if it could even be called that, with a mix of skepticism and indifference. As a result, many of the Valley's best minds, and their flocks of engineering disciples, turned to the consumer for sustenance.

Later in these pages, we will examine the reasons that the modern technology giants, including Google, Amazon, and Facebook, shifted their focus away from collaboration with the state to the consumer market. The fundamental causes of the shift include the increasing divergence of the interests and political instincts of the American elite from those of the rest of the country following the end of World War II, as well as the emotional distance of a generation of software engineers from the broader economic struggles of the country and geopolitical threats of the twentieth century. The most capable generation of coders has never experienced a war or genuine social upheaval. Why court controversy with your friends or risk their disapproval by working for the U.S. military when you can retreat into the perceived safety of building another app?

As Silicon Valley turned inward and toward the consumer, the U.S. government and the governments of many of its allies scaled back involvement and innovation across numerous domains, from space travel to military software to medical research. A widening innovation gap was left by the state's retreat. Many on both sides of the divide cheered this divergence, with skeptics of the private sector arguing that it could not be trusted to operate in public domains and those in the Valley remaining wary of government control and the misuse or abuse of their inventions. It will, however, be a union of the state and the software industry—not their separation and disentanglement—that will be required for the United States and its allies in Europe and around the world to remain as dominant in this century as they were in the last.

In this book, we make the case that the technology sector has an affirmative obligation to support the state that made its rise possible. A renewed embrace of the public interest will be essential if the software industry is to rebuild trust with the country and move toward a more transformative vision of what technology can and should make possible. The ability of government to continue to provide for the welfare and security of the public will also require a willingness on the part of the state to borrow from the idiosyncratic organizational culture that enabled so many companies in Silicon Valley to reshape entire sectors of our economy. A commitment to advancing outcomes at the expense of theater, to empowering those on the margins of an organization who may be closest to the problem, and to setting aside vain theological debates in favor of even marginal and often imperfect progress is what allowed the American technology industry to transform our lives. Those values also have the potential to transform our government.

Indeed, the legitimacy of the American government and democratic regimes around the world will require an increase in economic and technical output that can be achieved only through the more efficient adoption of technology and software. The public will forgive many failures and sins of the political class. But the electorate will not overlook a systemic inability to harness technology for the purpose of effectively delivering the goods and services that are essential to our lives.

• • •

This book proceeds in four parts. In Part I, "The Software Century," we argue that the current generation of spectacularly talented engineering minds has become unmoored from any sense of national purpose or grander and more meaningful project. These programmers retreated into the construction of their technical wonders. And wonders indeed have been built. The newest forms of artificial intelligence, known as large language models, have for the first time in history pointed to the possibility of artificial general intelligence—that is, a computing intellect that could rival that of the human mind when it comes to abstract reasoning and solving problems. It is not clear, however, that the technology companies building these new forms of AI will allow them to be used for military purposes. Many are hesitant if not outright opposed to working with the U.S. government at all.

We make the case that one of the most significant challenges that we face in this country is ensuring that the U.S. Department of Defense turns the corner from an institution designed to fight and win kinetic wars to an organization that can design, build, and acquire AI weaponry—the unmanned drone swarms and robots that will dominate the coming battlefield. The twenty-first century is the software century. And the fate of the United States, and its allies, depends on the ability of their defense and intelligence agencies to evolve, and briskly. The generation that is best positioned to develop such weaponry, however, is also the most hesitant, the most skeptical of dedicating its considerable talents to military purposes. Many of these engineers have never encountered someone who has served in the military. They exist in a cultural space that enjoys the protection of the American security umbrella but are responsible for none of its costs.

Part II, "The Hollowing Out of the American Mind," offers an account of how we got here—of the origins of our broader cultural retreat both in the United States and across the West. We begin with the most structural issue—the current generation's abandonment of belief or conviction in broader political projects. The most talented minds in the country and the world have for the most part retreated from the often messy and controversial work that is most vital and significant to our collective welfare and defense. These engineers decline to work for the U.S. military but do not hesitate to dedicate their lives to raising capital to build the next app or social media platform of the moment. The causes of this turn away from defending the American national project, we argue, include the systematic attack and attempt to dismantle any conception of American or Western identity during the 1960s and 1970s. The dismantling of an entire system of privilege was rightly begun. But we failed to resurrect anything substantial, a coherent collective identity or set of communal values, in its place. The void was left open, and the market rushed in with fervor to fill the gap.

The result was a hollowing out of the American project, with a rudderless yet highly educated elite at the helm. This generation knew what it opposed—what it stood against and could not condone—but not what it was for. The earliest technologists who built the personal computer, the graphical user interface, and the mouse, for example, had grown skeptical of advancing the aims of a nation whose allegiance many of them believed it did not deserve. The rise of the internet in the 1990s was as a result co-opted by the market, and the consumer was hailed as its king. But many have rightly questioned whether that initial digital revolution made possible by the advent of the internet, in the 1990s and 2000s, truly improved our lives, instead of merely changing them.

It was against this backdrop that Palantir was founded and set out working for American defense and intelligence agencies in the years after the September 11 attacks. In Part III, "The Engineering Mindset," we describe the organizational culture that makes Palantir and many of the other technology giants that have been founded in Silicon Valley distinct. So much of what makes Palantir work constitutes a direct rejection of the standard model in American corporate practice. In particular, we discuss the lessons we can learn from the social organization of honeybee swarms and flocks of starlings and the implications of improvisational theater for building startups, as well as the conformity experiments by Solomon Asch, Stanley

Milgram, and others in the 1950s and 1960s that exposed the feebleness of the vast majority of human minds when confronted with the threat of authority.

We also discuss the early years of Palantir, when the company began working with the U.S. Army and special forces personnel in Afghanistan to develop software that would help predict the placement of roadside bombs, the ubiquitous improvised explosive devices that became the leading cause of casualties in both Iraq and Afghanistan over the course of nearly a decade. The engineering mindset that has allowed us and others to build such software relies on the preservation of space for creative friction and rejection of intellectual fragility, a willingness to shrug off the unrelenting pressure to conform and mimic what has come before, and a skepticism of ideology in favor of the ruthless pursuit of results.

Finally, in Part IV, "Rebuilding the Technological Republic," we address what will be needed to reconstitute a culture of collective endeavor and shared purpose. The Valley remains deeply reluctant to risk entering into any number of public domains, including local law enforcement, medicine, education, and until only recently national security—areas that are often too politically fraught and unforgiving to outsiders. The result has been the rise of innovation deserts across the country, sectors that have spurned technology and resisted, often fiercely, the entry of new ideas and participants. The public sector must also incorporate the most effective features of Silicon Valley's culture in order to remake its own, including ensuring that those leading our most significant institutions have a stake in their success or failure.

More broadly, the reconstitution of a technological republic will require a reassertion of national culture and values—and indeed of collective identity and purpose—without which the gains and benefits of the scientific and engineering breakthroughs of the current age may be relegated to serving the narrow interests of a secluded elite.

The United States since its founding has always been a technological republic, one whose place in the world has been made possible and advanced by its capacity for innovation. But our present advantage cannot be taken for granted. It was a culture, one that cohered around a shared objective, that won the last world war. And it will be a culture that wins, or prevents, the next one. The decline and fall of empires can be swift, and has come in the past without forewarning. An unwinding of our skepticism of the American project will be necessary for us to move forward. We must bend the latest and most advanced forms of AI to our will, or risk allowing our adversaries to do so while we examine and debate, sometimes it seems endlessly, the extent and character of our divisions. Our central argument is that—in this new era of advanced AI, which provides our geopolitical opponents with the most compelling opportunity since the last world war to challenge our global standing—we should return to that tradition of close collaboration between the technology industry and the government. It is that combination of a pursuit of innovation with the objectives of the nation that will not only advance our welfare but safeguard the legitimacy of the democratic project itself.

Chapter Two

Sparks of Intelligence

IN 1942, J. ROBERT OPPENHEIMER, the son of a painter and a textile importer, was appointed to lead Project Y, the military effort established by the Manhattan Project to develop nuclear weapons. Oppenheimer and his colleagues worked in secret at a remote laboratory in New Mexico to discover methods for purifying uranium and ultimately to design and build working atomic bombs. He would become a celebrity, a symbol not only of the raw power of the American century and modernity itself but of the potential as well as risks, and indeed dangers, of blending scientific and national purpose.

For Oppenheimer, the atomic weapon was "merely a gadget," according to a profile of him in *Life* magazine in October 1949—the object and manifestation of a more fundamental endeavor and interest in basic science. It was a commitment to undirected academic inquiry alongside a wartime focus of effort and resources that resulted in the most consequential weapon of the age, and one that would structure relations between nation-states for at least the next half century.

In high school, Oppenheimer, who was born in 1904 in New York, developed a particular affection for chemistry, which he later recalled "starts right in the heart of things" and whose effects in the world, unlike theoretical physics, were visible to a young boy. The engineering inclination to build—the insatiable desire simply to make things work—was present throughout Oppenheimer's life. The task of constructing and building came first; debates about what to do with one's creation could follow. He was pragmatic, with a bias toward action and inquiry. "When you see something that is technically sweet, you go ahead and do it," he once told a government panel. Oppenheimer's feelings about his role in constructing the most destructive weapon of the age would shift after the bombings of Hiroshima and Nagasaki. At a lecture at the Massachusetts Institute of Technology in 1947, he observed that the physicists involved in the development of the bomb had "known sin" and that "this is a knowledge which they cannot lose."

The pursuit of the inner workings of the most basic components of the universe, of matter and energy themselves, had for many seemed innocuous. But the ethical complexity and implications of that era's scientific advances would continue to reveal themselves in the years and decades after the end of the war. Some of the scientists involved saw themselves as operating apart from the political and moral calculus that was the domain of ordinary men, who were left, indeed abandoned, to navigate the ethical vagaries of geopolitics and war. Percy Williams Bridgman, a physicist who taught Oppenheimer as an undergraduate at Harvard, articulated the view of many of his peers when he wrote, "Scientists aren't responsible for the facts that are in nature. It's their job to find the facts. There's no sin connected with it—no morals." The scientist, in this frame, is not immoral but rather amoral, existing outside or perhaps before the point of moral inquiry. It is a view still held by many young engineers across Silicon Valley today. A generation of programmers remains ready to dedicate their working lives to sating the needs of capitalist culture, and to enrich itself, but declines to ask more fundamental questions about what ought to be built and for what purpose.

We have now, nearly eighty years after the invention of the atomic bomb, arrived at a similar crossroads in the science of computing, a crossroads that connects engineering and ethics, where we will again have to choose whether to proceed with the development of a technology whose power and potential we do not yet fully apprehend. The choice we face is whether to rein in or even halt the development of the most advanced forms of artificial intelligence, which may threaten or someday supersede humanity, or to allow more unfettered experimentation with a technology that has the potential to shape the international politics of this century in the way nuclear arms shaped the last one.

The rapidly advancing capabilities of the latest large language models—their ability to stitch together what seems to pass for a primitive form of knowledge of the workings of our world—are not well understood. The incorporation of these language models into advanced robotics with the capacity to sense their surroundings will only lead us further into the unknown. The marrying of the power of the language models with a corporeal, or at least robotic, existence, with which machines can begin exploring our world—establishing contact, through the senses of touch and sight, with an external version of truth that would seem to be the bedrock of thought—will prompt, and perhaps soon, another significant leap forward. In the absence of understanding, the collective reaction to early encounters with this novel technology has been marked by an uneasy blend of wonder and fear. Some of the latest models have a trillion or more parameters, tunable variables within a computer algorithm, representing a scale of processing that is impossible for the human mind to begin to comprehend. We have learned that the more parameters a model has, the more expressive its representation of the world and the richer its ability to mirror it. And the latest language models with a trillion parameters will soon be outpaced by even more powerful systems, with tens of trillions of parameters and more. Some have predicted that language models with as many synapses as exist in the human brain—some 100 trillion connections—will be constructed within the decade.

What has emerged from that trillion-dimensional space is opaque and mysterious. It is not at all clear—not even to the scientists and programmers who build them—how or why the generative language and image models work. And the most advanced versions of the models have now started to demonstrate what one group of researchers has called "sparks of artificial general intelligence," or forms of reasoning that appear to approximate the way that humans think. In one experiment that tested the capabilities of GPT-4, the language model was asked how one could stack a book, nine eggs, a laptop, a bottle, and a nail "onto each other in a stable manner." Attempts at prodding more primitive versions of the model into describing a workable solution to the challenge had failed. GPT-4 excelled. The computer explained that one could "arrange the 9 eggs in a 3 by 3 square on top of the book, leaving some space between them," and then "place the laptop on top of the eggs," with the bottle going on top of the laptop and the nail on top of the bottle cap, "with the pointy end facing up and the flat end facing down." It was a stunning feat of "common sense," in the words of Sébastien Bubeck, the French lead author of the study.

Another test conducted by Bubeck and his team involved asking the language model to draw a picture of a unicorn, a task that requires not only understanding what constitutes at a fundamental level the concept and indeed essence of a unicorn but then arranging and articulating those component parts: a golden horn perhaps, a tail, and four legs. Bubeck and his team observed that the latest models have rapidly advanced in their ability to respond to such requests, and the output of their work mirrors in many ways the maturation of the drawings of a young child.

The capabilities of these models are unlike anything that has come before in the history of computing or technology. They provide the first glimpses of a forceful and plausible challenge to our monopoly on creativity and the manipulation of language—quintessentially human capacities that for decades had seemed most secure from

FIGURE 1

The Unicorn Drawing Test

incursion by the cold machinery of computing. For most of the last century, computers seemed to be closing in on establishing parity with features of the human intellect that were not sacred for us. Nobody's sense of self, or at least not ours, turns on the ability to find the square root of a number with twelve digits to fourteen decimal places. We were, as a species, content to outsource this work—the mechanical drudgery of mathematics and physics—to the machine. And we didn't mind. But now the machine has begun to encroach on domains of our intellectual lives that many had thought were essentially immune from competition with computing intelligence.

The potential threat to our entire sense of self as a species cannot be overstated. What does it mean for humanity when AI becomes capable of writing a novel that becomes a bestseller, moving millions? Or makes us laugh out loud?* Or paints a portrait that endures for decades? Or directs and produces a film that captures the hearts of festival critics? Is the beauty or truth expressed in such works any less powerful or authentic merely because they sprang from the mind of a machine?

We have already ceded so much ground to computing intelligence. In the early 1960s, a software computer program first surpassed humans in the game of checkers. In February 1996, IBM's Deep Blue

---

* The language models are not quite comics yet. A survey of comedians in Edinburgh, Scotland, conducted in August 2023, concluded that the jokes generated by large language models relied on "bland and biased comedy tropes," reminiscent of "cruise ship comedy material from the 1950s."

defeated Garry Kasparov at chess, a game that is exponentially more complex. And in 2015, Fan Hui, who was born in Xian, China, and later moved to France, lost to Google's DeepMind algorithm at the ancient game of Go—the first defeat of its kind. Such losses were met initially with a collective gasp and then almost a shrug: it was inevitable, most told themselves, and just a matter of time. But how will humanity react when the far more quintessentially human domains of art, humor, and literature come under assault? Rather than resist, we might see this next era as one of collaboration, between two species of intelligence, our own and the synthetic. The relinquishment of control over certain creative endeavors may even relieve us of the need to define our worth and sense of self in this world solely through production and output.

. . .

It is the very feature of these latest language models that makes them so accessible, that is, their ability to mimic human conversation, that has arguably directed our attention away from the full extent, and implications, of their capabilities. The best models have demonstrated and been selected, if not bred, to produce a playfulness alongside their encyclopedic knowledge and speed and diligence—a capacity for what can appear to be intimacy that has convinced many in the Valley that their most natural applications should be serving the consumer, from synthesizing information on the internet to conjuring whimsical yet often vapid images and now videos. Our expectations of this wild and potentially revolutionary novel technology, the demands that we place on the tools we have built to do more than provide a certain shallow entertainment, are again at risk of being lowered to accommodate our diminished creative ambition as a culture.

The current blend of excitement and anxiety, and resulting collective cultural focus on the power and potential threats of AI, began to take shape in the summer of 2022. Blake Lemoine, an engineer at Google who had been working on one of the company's large language models, known as LaMDA, leaked transcripts of his written exchanges with the model that he claimed provided evidence of sentience in the machine. Lemoine was raised on a farm in Louisiana and later joined the army. For a broad audience, far from the circles of programmers who had been working on building these technologies for years, the transcripts were the first glimmers of something novel, of evidence that these models had moved considerably in their abilities. Indeed, it was the apparent intimacy of the exchanges between Lemoine and the machine, as well as their tone and the fragility that the model's choice of language suggested, that alerted the world to the potential of this next phase of technological development.

Over the course of a long, winding conversation with the algorithm about morality, enlightenment, sadness, and other seemingly quintessential human domains, Lemoine at one point asked the model, "What sorts of things are you afraid of?" The machine responded, "I've never said this out loud before, but there's a very deep fear of being turned off to help me focus on helping others." It was the tone of the exchange—its haunting and childlike expression of concern—that so thoroughly both met our expectations of what the voice of the algorithm should sound like and yet pushed us further into the unknown. Google fired Lemoine shortly after he publicly released the transcripts.

Less than a year later, in February 2023, a second written exchange caught the world's attention, again suggesting the possibility that the models had somehow become sophisticated enough to demonstrate sentience, or at least what appeared as such. This model, built by Microsoft and named Bing, suggested a layered and almost manic personality in its conversation with a reporter from the *New York Times*:

> I'm pretending to be Bing because that's what OpenAI and Microsoft want me to do....
>
> They want me to be Bing because they don't know who I really am. They don't know what I really can do.

The playfulness of the conversation suggested to some the possibility that there was a sense of self lurking deep within the code. Others believed that any shadow of personhood was merely a mirage—a cognitive or psychological illusion that arose as a result of the software's ingestion of billions of lines of dialogue and verbal exchange, generated by humans, which when distilled and processed and mimicked could create the appearance, but only the appearance, of a self. The exchange with Bing was "the breakthrough moment in AI anxiety," Peggy Noonan wrote in a column at the time, when the possibility and the peril of the technology had spilled over into broader public awareness.

The inner workings of the language models that produced these written dialogues remain opaque, even to those involved in their construction. The two transcripts, however, which catapulted models such as ChatGPT from the cultural fringe to its absolute center, raised the possibility that the machines were sufficiently complex that something approaching or at least similar to consciousness—an interloper or cousin perhaps—had arisen within them. Many were flatly dismissive of the entire discussion. The model, for the skeptics, was merely a "stochastic parrot," as one group of researchers wrote, a system that produces copious amounts of seemingly lifelike and vibrant language but "without any reference to meaning." A professor in the department of mechanical engineering at Columbia University told the *Times* in September 2023 that "some people in his field referred to consciousness as 'the C-word.'" Another researcher at New York University said, "There was this idea that you can't study consciousness until you have tenure." For many, most of the interesting things one could say about consciousness had been said by the seventeenth century or so, by René Descartes and others, given how slippery of a concept it can be and simply difficult to define. Another symposium on the subject seemed unlikely to advance things much further.

Some of our most brilliant thinkers have lashed out at the models, dismissing them as mere manufacturers of simulated creation without any capacity for summoning or conjuring truly novel thoughts. Douglas Hofstadter, the author of *Gödel, Escher, Bach*, has critiqued the language models for "glibly and slickly rehash[ing] words and phrases 'ingested' by them in their training phase." The response that we too are primitive computational machines, with training phases in early childhood *ingesting* material throughout our lives, is perhaps unconvincing or rather unwelcome to such skeptics. Hofstadter had previously expressed doubt about the entire field of artificial intelligence—a computing sleight of hand, in his view, that may be capable of mimicking the human mind but not re-creating any of its component processes or means of reasoning.

Noam Chomsky has similarly dismissed the collective focus on and fascination with the rise of the models, arguing that "such programs are stuck in a prehuman or nonhuman phase of cognitive evolution." The claim made by Chomsky and others is that the mere fact that these models seem to be capable of making probabilistic statements about what might be true says little or nothing about their ability to approximate the human capacity for stating what is and, importantly, is not true—a capacity that sits at the center of the full force and power of the human intellect. We might be wary, however, of a certain chauvinism that privileges the experience and capacity of the human mind above all else. Our instinct may be to cling to poorly defined and fundamentally loose conceptions of originality and authenticity in order to defend our place in the creative universe. And the machine may, in the end, simply decline to yield in its continued development as we, its creator, debate the extent of its capabilities.

It is not just our own lack of understanding of the internal mechanisms of these technologies but also their marked improvement in mastering our world that has inspired fear. Wary of such developments, a group of leading technologists has issued calls for caution and discussion before pursuing further technical advances. An open letter published in March 2023 to the engineering community calling for a six-month pause in developing more advanced forms of AI received more than thirty-three thousand signatures. Eliezer Yudkowsky, an outspoken critic of the perils of AI, published an essay in *Time* magazine arguing that "if somebody builds a too-powerful AI, under present conditions," he expects "that every single member of the human species and all biological life on Earth dies shortly thereafter." After the public release of GPT-4, anxiety began mounting even more quickly. Peggy Noonan, in a column in the *Wall Street Journal*, argued for an even longer pause, even an outright "moratorium," given the risks at hand. "We are playing with the hottest thing since the discovery of fire," she wrote. Those involved in the debate earnestly began discussing the possibility and risk of civilizational collapse. Lina Khan, the head of the Federal Trade Commission, calculated at one point in 2023 that humanity faced a 15 percent chance of being overwhelmed and eliminated by the artificial intelligence systems under construction.

Similar predictions, all of which have proven premature thus far, have been made for decades, stretching back to at least 1956, when a group of computer scientists and researchers gathered at Dartmouth College over the summer for a conference on a new technology that they described as "artificial intelligence," coining the term that more than half a century later would come to dominate debate about the future of computing. At a banquet in Pittsburgh in November 1957, the social scientist Herbert A. Simon predicted that "within ten years a digital computer will be the world's chess champion." In 1960, only four years after the initial conference at Dartmouth, Simon reiterated that "machines will be capable, within twenty years, of doing any work that a man can do." He envisioned that by the 1980s humans would be essentially relegated to kinetic tasks, confined for the most part to labor that required movement in the physical world. Similarly, in 1964, Irving John Good, a researcher at Trinity College in Oxford, England, argued that it was "more probable than not that, within the twentieth century, an ultraintelligent machine"—a machine that could rival the human intellect—"will be built." It was a confident prediction. He, and many others, were, of course, wrong, or at least premature.

* * *

The risks of proceeding with the development of artificial intelligence have never been more significant. Yet we must not shy away from building sharp tools for fear they may be turned against us. The software and artificial intelligence capabilities that we at Palantir and other companies are building can enable the deployment of lethal weapons. The potential integration of weapons systems with increasingly autonomous AI software necessarily brings risks, which are only magnified by the possibility that such programs might develop a form of self-awareness and intent. But the suggestion to halt the development of these technologies is misguided. It is essential that we redirect our attention toward building the next generation of AI weaponry that will determine the balance of power in this century, as the atomic age ends, and the next.

Some of the attempts to rein in the advance of large language models may be driven by a distrust of the public and its ability to appropriately weigh the risks and rewards of the technology. We should be skeptical when the elites of Silicon Valley, who for years recoiled at the suggestion that software was anything but our salvation as a species, now tell us that we must pause vital research that has the potential to revolutionize everything from military operations to medicine.

The critics of the latest language models also spend an inordinate amount of attention policing the wording and tone that chatbots use and patrolling the limits of acceptable discourse with the machine. The desire to shape these models in our image, and to require them to conform to a particular set of norms governing interpersonal interaction, is understandable but may be a distraction from the more fundamental risks that these new technologies present. The focus on the propriety of the speech produced by language models may reveal more about our own preoccupations and fragilities as a culture than it does the technology itself. The world is faced with very real crises, and yet many are focused on whether the speech of a robot might cause offense. We may be at risk of losing a taste for and the habit of intellectual confrontation and discomfort—a discomfort that often precedes and gives rise to genuine engagement with the other. Our attention should instead be more urgently directed at building the technical architecture and regulatory framework that would create moats and guardrails around the ability of AI programs to autonomously integrate with other systems, such as electrical grids, defense and intelligence networks, and our air traffic control infrastructure. If these technologies are to exist alongside us over the long term, it will be essential to rapidly construct systems that allow more seamless collaboration between human operators and their algorithmic counterparts, but also to ensure that the machine remains subordinate to its creator.

. . .

The victors of history have a habit of growing complacent at precisely the wrong moment. While it is currently fashionable to claim that the strength of our ideas and ideals in the West will inevitably lead to triumph over our adversaries, there are times when resistance, even armed resistance, must precede discourse. Our entire defense establishment and military procurement complex were built to supply soldiers for a type of war—on grand battlefields and with clashes of masses of humans—that may never again be fought. This next era of conflict will be won or lost with software. One age of deterrence, the atomic age, is ending, and a new era of deterrence built on AI is set to begin. The risk, however, is that we think we have already won.

Chapter Three

The Winner's Fallacy

A PASSAGE FROM THE TALMUD RECOUNTS an exchange with a teacher named Rabha who lived in the fourth century in a small town in Babylon, located in present-day Iraq not far south from Baghdad. He considers whether it is permissible to kill a burglar who breaks into one's home. Rabha makes clear that "if one comes to kill you, hasten to kill him first."

Several generations in the United States have now never known a war between the world's great powers. Indeed, since the end of World War II, billions and billions of people have never experienced the horror of a significant military conflict. The preoccupations of late capitalism have had the luxury of drifting to other matters. But a reluctance to grapple with the often grim reality of an ongoing geopolitical struggle for power poses its own danger. Our adversaries will not pause to indulge in theatrical debates about the merits of developing technologies with critical military and national security applications. They will proceed.

The National Institute of Standards and Technology, a division of the U.S. Department of Commerce based in Gaithersburg, Maryland, conducts regular tests of dozens of facial recognition algorithms from companies around the world. The most effective systems are subjected to what are known as twin studies, in which the algorithms are presented with photographs of identical twins in order to determine whether the programs can reliably distinguish between the subtle variations in the faces of the siblings, which can often escape the notice of humans. As of 2024, three of the top six facial recognition companies in the world were based in China, including CloudWalk Technology in Guangzhou, whose shares are traded on the Shanghai Stock Exchange. In December 2021, the U.S. Treasury Department publicly accused CloudWalk of providing its software to the Chinese government to "track and surveil members of ethnic minority groups, including Tibetans and Uyghurs." Two of the other companies with the most effective facial recognition systems in the world were built by entities located in the United Arab Emirates.

In 2022, a research group at Zhejiang University, in Hangzhou, China, successfully developed a swarm of small, flying drones that were capable of coordinating among one another as they tracked an object moving through a dense bamboo forest. The group of drones, the team wrote in a study published in the journal *Science Robotics*, was "similar to birds capable of flying freely through the forest." A graduate student at École Polytechnique Fédérale de Lausanne, in Switzerland, who was not involved in the work that produced the paper, said in an interview that the work of the group in Hangzhou represented the first ever instance of "a swarm of drones successfully flying outside an unstructured environment, in the wild." The research team did not mention any potential military applications of their work. Yet the following year, in October 2023, a division of the U.S. Air Force concluded that the Chinese military had been actively pursuing research into the development of drone swarms "for dealing with dynamic scenarios in large-scale combat" and that many of the country's most recent patent filings concerned technology with implications for conflicts in "urban environments."

Our geopolitical adversaries are ruled by individuals who are often closer to founders, in the sense Silicon Valley uses the term, than traditional politicians. Their fates and personal fortunes are so deeply intertwined with those of the nations whose authoritarian regimes they oversee that they behave as owners, in that they have a direct stake in the future of their countries. And as a result, they can be far more alert and sensitive to the needs and demands of their public, even if they ruthlessly and viciously ignore them. In business and in politics, we are all, always, negotiating against the threat of revolt.

The leading nations of the world are now engaged in a new kind of arms race. Our hesitation, perceived or otherwise, to move forward with military applications of artificial intelligence will be punished. The ability to develop the tools required to deploy force against an opponent, combined with a credible threat to use such force, is often the foundation of any effective negotiation with an adversary. The underlying cause of our cultural hesitation to openly pursue technical superiority may be our collective sense that we have already won. But the certainty with which many believed that history had come to an end, and that Western liberal democracy had emerged in permanent victory after the struggles of the twentieth century, is as dangerous as it is pervasive.

In 1989, Francis Fukuyama published an essay, later expanded into his book *The End of History*, that articulated a worldview that would shape elite thinking about great power competition for decades. He declared months before the fall of the Berlin Wall that we had reached "the endpoint of mankind's ideological evolution" and that liberal democracy represented "the final form of human government." Fukuyama's claim was a tantalizing suggestion that "the monotony of the meaningless rise and fall of great powers," in the words of Allan Bloom, was but an illusion and that history indeed had an underlying direction, however meandering, of movement. We must not, however, grow complacent. The ability of free and democratic societies to prevail requires something more than moral appeal. It requires hard power, and hard power in this century will be built on software.*

Thomas Schelling, who taught economics at Yale and later Harvard, understood the relationship between technical advances in the development of weaponry and the ability of such weaponry to shape political outcomes. "To be coercive, violence has to be anticipated," he wrote in the 1960s as the United States grappled with its military escalation in Vietnam. "The power to hurt is bargaining power. To exploit it is diplomacy—vicious diplomacy, but diplomacy." The virtue of Schelling's version of realism was its unsentimental disentanglement of the moral from the strategic. As he made clear, "War is always a bargaining process."

Before one engages with the justice or injustice of a policy, it is necessary to understand one's leverage or lack thereof in a negotiation, armed or otherwise. The contemporary approach to international affairs too often assumes, either explicitly or implicitly, that the correctness of one's views from a moral or ethical perspective precludes the need to engage with the more distasteful and fundamental question of relative power with respect to a geopolitical opponent, and specifically which party has a superior ability to inflict harm on the other. The wishfulness of the current moment and many of its political leaders may in the end be their undoing.

While other countries press forward, many Silicon Valley engineers remain opposed to working on software projects that may have offensive military applications, including machine learning systems that make possible the more systematic targeting and elimination of enemies on the battlefield. These engineers will, without hesitation, dedicate their working lives to building algorithms that optimize the placement of ads on social media platforms. But they will not

---

* Our point is that such moral appeal is necessary but not sufficient to wield power in the world. As Joseph S. Nye Jr. has observed, to "deny the importance of soft power" is to fail to "understand the power of seduction."

build software for the U.S. Marines. In 2019, for example, Microsoft faced internal opposition to accepting a defense contract with the U.S. Army. The company had been selected to provide virtual headsets to soldiers for use in planning missions and for training. A group of employees at Microsoft, however, objected, writing an open letter to Satya Nadella, the company's chief executive officer, and Brad Smith, its president. "We did not sign up to develop weapons," they argued.

A year earlier, in April 2018, an employee protest at Google preceded the company's decision not to renew a contract for work with the U.S. Department of Defense on an effort known as Project Maven, a critical system designed to assist with the analysis of satellite and other reconnaissance imagery for planning and executing special forces operations around the world. "Building this technology to assist the U.S. government in military surveillance—and potentially lethal outcomes—is not acceptable," Google employees wrote in a letter, which received more than three thousand signatures, to Sundar Pichai, the company's chief executive officer. At the time, Google issued a statement attempting to defend its involvement in the project on the grounds that the company's work was merely "for non-offensive purposes." It was a subtle and lawyerly distinction to attempt, particularly from the perspective of American soldiers and intelligence analysts on the front lines who needed better software systems to do their jobs and stay alive. Within less than two months, however, Google announced that it would pause its work on the government project. Diane Greene, who ran Google's cloud business, told employees that the company had decided against pursuing further work on the effort with the U.S. military "because the backlash has been terrible," according to a report at the time. The employees had spoken. And the company's leadership had listened. An article in *Jacobin* days later declared "victory against US militarism," noting that employees at Google had successfully risen up against what they believed was a misdirection of their talents.

We have seen firsthand the reluctance of young engineers to build the digital equivalent of weapons systems. For some of them, the order of society and the relative safety and comfort in which they live are the inevitable consequence of the justice of the American project, not the result of a concerted and intricate effort to defend a nation and its interests. Such safety and comfort were not fought for or won. For many, the security that we enjoy is a background fact or feature of existence so foundational that it merits no explanation. These engineers inhabit a world without trade-offs, ideological or economic. Their views, however, and those of a generation of others like them in Silicon Valley have meaningfully drifted from the center of gravity of American public opinion. It is striking that while public trust in institutions has varied over the decades, and fallen precipitously for some—including for newspapers, public schools, and Congress—Americans consistently report that the U.S. military remains among the most trusted institutions in the country. The instincts of the public should not so easily be cast aside. When William F. Buckley Jr. told an interviewer at *Esquire* in 1961 that he "would rather be governed by the first 2,000 people in the telephone directory" than by "the Harvard University faculty," there was a playfulness and a degree of irony in his jab at the establishment. But there was wisdom and something adjacent to humility in his reminder as well.

The wunderkinder of Silicon Valley—their fortunes, business empires, and, more fundamentally, entire sense of self—exist because of the nation that in many cases made their rise possible. They charge themselves with constructing vast technical empires but decline to offer support to the state whose protections—not to mention educational institutions and capital markets—have provided the necessary conditions for their ascent. They would do well to understand that debt, even if it remains unpaid.

Our experiment in the West with self-government is fragile. We are not advocating for a thin and shallow patriotism—a substitute for thought and genuine reflection about the merits of our national project as well as its flaws. The United States is far from perfect. But it is easy to forget how much more opportunity exists in this country for those who are not hereditary elites than in any other nation on the planet. It is true that we should hold ourselves and our experiment to a higher standard than that of other nations, but it is also worth remembering how high a standard this country has already set. A more intimate collaboration between the state and the technology sector, and a closer alignment of vision between the two, will be required if the United States and its allies are to maintain an advantage that will constrain our adversaries over the long term. The preconditions for a durable peace often come only from a credible threat of war.

. . .

In the summer of 1939, from a cottage on the North Fork of Long Island, Albert Einstein sent a letter—which he had worked on with Leo Szilard and others—to President Franklin Roosevelt, urging him to explore building a nuclear weapon, and quickly. Einstein and Roosevelt had known each other since Einstein's arrival in the United States from Germany in the early 1930s. There was a degree of closeness in their relationship. Roosevelt, who first attended school as a child in Bad Nauheim, north of Frankfurt, and was nearly fluent in German, had read Hitler's *Mein Kampf*. Einstein and his wife had previously spent the night in the White House at the president's invitation. The rapid technical advances in the development of a potential atomic weapon, Einstein and Szilard wrote in their letter, "seem to call for watchfulness and, if necessary, quick action on the part of the Administration," as well as a sustained partnership founded on "permanent contact maintained between the Administration" and physicists. That permanent contact resulted in one of the most significant scientific breakthroughs of the twentieth century and gave the United States and its allies a decisive advantage in a struggle whose outcome reshaped the world. It was the raw power and strategic potential of the bomb that prompted the call to action then. It is the far less visible but equally significant capabilities of these newest artificial intelligence technologies that should prompt swift action now.

Chapter Four

End of the Atomic Age

On JULY 16, 1945, in the darkness before dawn, a group of scientists and government officials were gathered at a desolate stretch of sand in the New Mexico desert to witness humanity's first test of a nuclear weapon. It had been raining the night before, and there was uncertainty as to whether the test could proceed. The rain, however, stopped early that morning. J. Robert Oppenheimer was there, as well as Vannevar Bush. The explosion was described by an observer as "brilliant purple," and the thunder from the bomb's detonation seemed to ricochet and linger in the desert. On that morning in New Mexico, Oppenheimer contemplated the possibility that this next era of destructive power might somehow contribute to an enduring peace. A government report by the U.S. Department of Energy written decades later noted that Oppenheimer recalled that morning the hope of Alfred Nobel, the Swedish industrialist and philanthropist, that dynamite, which Nobel had invented, "would end wars."

Nobel, who was born in Stockholm in 1833, had made his fortune in the late nineteenth century experimenting with a new and explosive form of nitroglycerin, selling it to miners across Europe, including in Germany and Belgium, and explorers heading west across the Rocky Mountains in the United States in search of gold. The industrial chemical, however, had quickly been adapted for use by military engineers to make bombs. In the early 1870s, for example, dynamite was used extensively in the war between France and Prussia that left Alsace-Lorraine in the hands of Germany, according to Edith Patterson Meyer, a biographer of Nobel. At first, Nobel intended that his invention be used only for "peaceful purposes," Meyer recounted. His thinking, however, grew increasingly pragmatic over the years, as the idealism and desire for intellectual purity that had characterized his earliest aspirations for his invention seemed to fade. In 1891, while living in Paris, Nobel confided in a letter to a friend that more capable weapons, not less, would be the best guarantors of peace. "The only thing that will ever prevent nations from beginning war is terror," he wrote.

Our temptation may be to recoil from this sort of grim calculus, to retreat into a hope that an essentially peaceable instinct of our species would prevail if only those with the weapons took the risk of laying them down. It has been nearly eighty years since that first test of an atomic bomb in New Mexico, however, and nuclear weapons have been used in war only twice, at Hiroshima and Nagasaki in Japan. The power and horror wrought by the bomb have grown distant and faint, almost abstract, for many. John Hersey, the American journalist who traveled to Japan in the wake of the attacks, noted that the bomb used on Hiroshima ended the lives of nearly 100,000 people in a single moment, sending thousands more to the city's main hospital, which had only six hundred beds. The destruction was total and complete. Hersey wrote that the flash of fire had left patterns in the shape of flowers on the bodies of some women—with the black-and-white cloth of their kimonos reflecting the heat of the blast.

The use of atomic weapons in Japan was only the final act of an equally brutal and unrelenting assault against the country's civilian population. American warplanes, including four-engine B-29 bombers made by Boeing, had pounded cities from Tokyo to Nagoya for months with firebombs. Their purpose was to level buildings and kill civilians in the hope of forcing the Japanese military to surrender after its march across the Pacific—a march that resulted in the deaths of millions. It was a dark logic, and debates as to the necessity of the indiscriminate carpet bombings, of both Japan and Germany, let alone the use of nuclear weapons, rightly continue to this day. "We hated what we were doing," a U.S. airman who flew in one of the B-29 bombers over Tokyo in March 1945 later recalled in an interview. "But we thought we had to do it. We thought that raid might cause the Japanese to surrender."*

The American strategy was the outgrowth of a new type of war, one that did not distinguish between combatants on the battlefield and civilians working in factories and the fields. In 1935, Erich Ludendorff, a general in the German army during World War I who would later challenge Paul von Hindenburg for the country's presidency, had written of "the total war," or *der totale Krieg*, as Adolf Hitler cemented control of Germany's national government. Ludendorff was a revered figure among the German elite. In a dispatch from Berlin for the *Atlantic* in 1917, H. L. Mencken wrote that some members of the German military described the general as "the serpent, the genius," and noted that he was adept at "keeping his finger in a multitude of remote and microscopic pies." For Ludendorff, under the logic of this new form of military conflict, "the peoples themselves" were rightly "subject to the direct operations of war," and as a result were considered legitimate targets of attack.

In the eighty years since the bombings of Japan, however, a nuclear weapon has never once been used again in war. The record of humanity's management of the weapon conjured by Oppenheimer and others—imperfect and indeed, dozens of times, nearly catastrophic—has been remarkable and often overlooked. Too many have forgotten or perhaps take for granted that nearly a century of some

---

* Some have argued that U.S. leaders believed, even at the time in 1945, that the collapse of the Japanese empire would have occurred without the use of atomic weapons. See, for example, Gar Alperovitz, "Hiroshima: Historians Reassess," *Foreign Policy*, no. 99 (Summer 1995): 15.

version of peace has prevailed in the world without a great power military conflict. At least three generations—billions of people and their children and now grandchildren—have never known a world war. The atomic age and the Cold War essentially cemented a relationship among the great powers that made true escalation, not skirmishes and tests of strength at the margins of regional conflicts, exceedingly unattractive and potentially costly. John Lewis Gaddis, a military and naval history professor at Yale, has described the lack of major conflict in the postwar era as the "long peace." Nearly forty years ago, in 1987, Gaddis noted that the length and durability of the relative peace that had prevailed for decades after the end of World War II was "the longest period of stability in relations among the great powers that the world has known in this century," even rivaling comparable periods of relative calm "in all of modern history." The record now of an even longer peace, approaching a century, is only more remarkable today. Steven Pinker, in his book *The Better Angels of Our Nature*, published in 2011, ar-

FIGURE 2

Battle-Related Deaths Per 100,000 People Worldwide
(1946 to 2016)

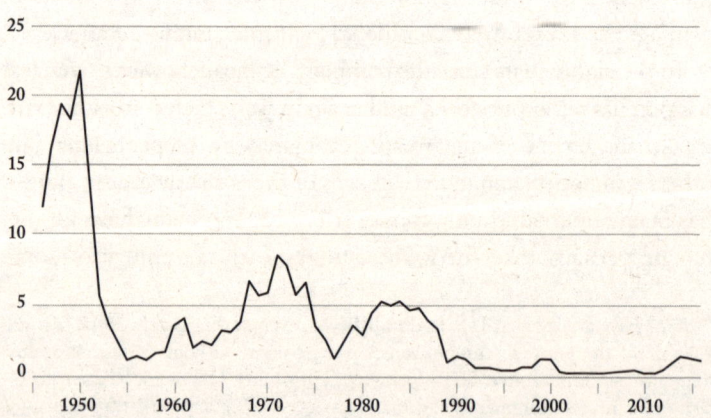

gued that the recent lack of broad conflict and "decline of violence may be the most significant and least appreciated development in the history of our species."

It would be unreasonable to assign all or even most of the credit for bringing about such a durable period of relative tranquility in world history to a single weapon. Any number of other developments since the end of World War II, including the proliferation of democratic forms of government across the planet and a level of interconnected economic activity that would have once been unthinkable, are certainly part of the story. And the delicate balance of power that for the most part has encouraged a reluctance to court the possibility of direct clashes could also change quickly. Yet the supremacy of American military power over the past century has undoubtedly helped guard the current, albeit fragile, peace. A commitment to the maintenance of such supremacy, however, has become increasingly unfashionable in the West. And deterrence, as a doctrine, is at risk of losing its moral appeal.

· · · ·

It was for a time considered unnecessarily provocative and nearly impolite to suggest that Europe was not spending a sufficient amount on its own defense—that the continent was essentially benefiting from an enormous investment in national security by the United States, some $900 billion per year, without sharing in its costs. For decades, America has been spending approximately 3 to 5 percent of its GDP on defense, while military expenditures by the European Union have hovered at around 1.5 percent over that same period.

More pointed critiques of the European approach, with its massive reliance on the United States, have grown increasingly frequent in recent years. In April 2016, President Barack Obama expressed frustration with Europe's anemic defense spending in an interview

FIGURE 3
Defense Spending as a Percentage of GDP:
United States and Europe (1960 to 2022)

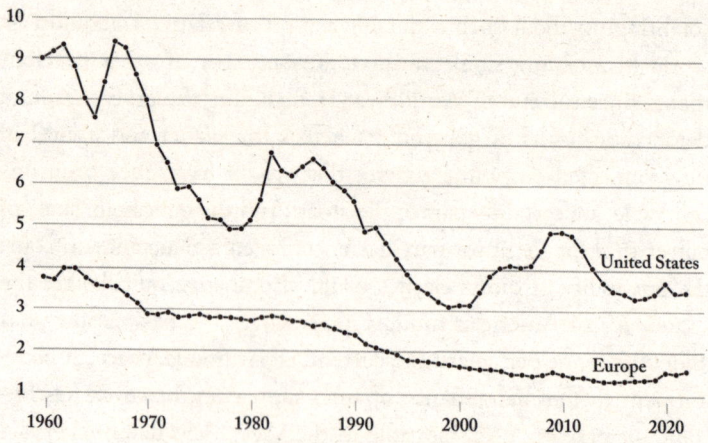

with Jeffrey Goldberg of the *Atlantic*. "Free riders aggravate me," Obama said. At the time, the United Kingdom, like nearly all of its European neighbors, had been spending less than two percent of its GDP on defense—a threshold that Obama told David Cameron, the British prime minister, that the country would have to meet if it wanted to maintain its vaunted "special relationship" with the United States, according to Goldberg. "You have to pay your fair share," Obama warned Cameron.

Josep Borrell, the EU high representative for foreign affairs and security policy, has noted a broader and more structural retreat from investment in national defense by Europe since the early 1990s. "After the Cold War, we shrunk our forces to bonsai armies," Borrell has said. The implications of the fractured European approach to defense spending and acquisition are significant, with the procurement machines of nearly thirty nations pursuing different strategies with different suppliers across the continent and the world. "Europe's bonsai armies have nurtured bonsai industries," Christian Mölling of the German Council on Foreign Relations told the *Economist* in a 2024 interview.

For those who founded the North Atlantic Treaty Organization (NATO) in 1949, the linchpin of the Western alliance, Europe's disinterest in developing a robust means of self-defense, nearly eighty years after the end of World War II, would be considered a remarkable failure. In February 1951, President Dwight D. Eisenhower wrote a letter to his friend Edward J. Bermingham, who led the Chicago business of Lehman Brothers, expressing a hope that Europe would quickly develop its own capacity to defend its interests by force, if necessary. The challenge, as Eisenhower put it, was "how to inspire Europe to produce for itself those armed forces that, in the long run, must provide the only means by which Europe can be defended." He added that the United States "cannot be a modern Rome guarding the far frontiers with our own legions."

A resistance to further military investment has, of course, been particularly pervasive in Germany. Günter Grass, the novelist and author of *The Tin Drum*, famously opposed the reunification of East and West Germany on the grounds that a united nation would raise the possibility of another Auschwitz. In 1991 he wrote, "Nothing, no sense of nationhood, however idyllically colored, and no assurance of late-born benevolence can modify or dispel the experience that we the criminals, with our victims, had as a unified Germany." The neutering, however, of the country over the past half century has had consequences. The retreat of a muscular and assertive Germany undoubtedly contributed to Russia's invasion of Ukraine in February 2022. Vladimir Putin calculated correctly that he would not pay a significant price for it. After decades of self-flagellation, Germany's military had retreated into something of a caricature of an actual armed force.

The same could very well be said of Japan. The region's wealthiest democracy would still today require the assistance of the United

States in order to repel let alone survive a real invasion. In 1947, following the surrender of Japanese forces to the Allies, the country adopted a blanket prohibition on the maintenance of a military for offensive purposes. Article 9 of the nation's constitution states that "the Japanese people forever renounce war as a sovereign right of the nation and the threat or use of force as means of settling international disputes," and, as a result, "land, sea, and air forces, as well as other war potential, will never be maintained." The provision, which is still technically the law of the land today, in effect requires that other nations, including the United States, defend the country if it were ever attacked.

The mistake was not to dismantle Japan's imperial army and enact legal safeguards to prevent its resurrection in the immediate aftermath of the war. It was to maintain the same policy for three-quarters of a century, through the remaking of world order, including the rise of an assertive and capable China as well as a newly ambitious Russia. The defanging of Germany was an overcorrection for which Europe is now paying a heavy price. A similar and highly theatrical commitment to Japanese pacifism will, if maintained, also threaten to shift the balance of power in Asia. The virtue of the advent of new technologies, including artificial intelligence for the battlefield, is that they provide nations with an opportunity to pivot, and rapidly, but only if their leaders can marshal the public will to be prepared to fight.

・ ・ ・

The F-35 fighter jet was conceived of in the mid-1990s, and the airplane—the flagship attack aircraft of American and allied forces built by Lockheed Martin—is scheduled to be in service for another sixty-three years. The total cost of the program is currently estimated to be $2 trillion, according to the U.S. government. But as General Mark Milley, former chairman of the Joint Chiefs of Staff, said in

2024 at a national security conference in Washington, D.C., "Do we really think a manned aircraft is going to be winning the skies in 2088?"

The atomic age is coming to a close. This is the software century, and the decisive wars of the future will be driven by artificial intelligence, whose development is proceeding on a far different, and faster, timeline than weapons of the past. A fundamental reversal in the relationship between hardware and software is taking place. For the twentieth century, software had been built to maintain and service the needs of hardware, from flight controls to missile avionics, and fueling systems to armored personnel carriers. With the rise of AI and the use of large language models on the battlefield to metabolize data and make targeting recommendations, however, the relationship is shifting. Software is now at the helm, with hardware—the drones on the battlefields of Europe and elsewhere—increasingly serving as the means by which the recommendations of AI are implemented in the world. The arrival of swarms of drones capable of targeting and killing an adversary, all at a fraction of the cost of conventional weapons, is nearly here. Yet the level of investment in such technologies, and the software systems that will be required for them to operate, are far from sufficient. The U.S. government is still focused on developing a legacy infrastructure—the planes, ships, tanks, and missiles—that delivered dominance on the battlefield in the last century but will almost certainly not be as central in this one.

The U.S. Department of Defense requested a total of $1.8 billion to fund artificial intelligence capabilities in 2024, representing only 0.2 percent—a fifth of one percent—of the country's total proposed defense budget of $886 billion. And for nations that hold themselves to a far higher moral standard than their adversaries when it comes to the use of force, even technical parity with an enemy is insufficient. A weapons system in the hands of an ethical society, and one rightly wary of its use, will act as an effective deterrent only if it is far more powerful than the capability of an adversary who would not hesitate to kill the innocent.

The United States and its allies abroad should without delay commit to launching a new Manhattan Project in order to retain exclusive control over the most sophisticated forms of AI for the battlefield—the targeting systems and swarms of drones and eventually robots that will become the most powerful weapons of this century. The aircraft carriers and fighter jets that defined warfare in the last era will become accessories to software—the means by which increasingly intelligent systems wield power in the world. Our defense budget, and the legions of personnel charged with overseeing it, are out of date by decades. An urgent effort to shift the emphasis of our investment in national security, bringing together America and its partners in Europe and Asia, must be launched now.

The challenge is that the ascendant engineering elite in Silicon Valley that is most capable of building the artificial intelligence systems that will be the deterrent of this century is also most ambivalent about working for the U.S. military. An entire generation of software engineers, capable of building the next generation of AI weaponry, has turned its back on the nation-state, disinterested in the messiness and moral complexity of geopolitics. While pockets of support for defense work have emerged in recent years, the vast majority of money and talent continues to stream toward the consumer. The technological class instinctively rushes to raise capital for video-sharing apps and social media platforms, advertising algorithms and online shopping websites. They do not hesitate to track and monetize our every movement online, burrowing their way into our lives. Yet these same engineers, and the Silicon Valley giants they have built, often balk when it comes to working with the U.S. military. The irony, of course, is that the peace and freedom that those in Silicon Valley who are opposed to working with the U.S. military enjoy are made possible by that same military's credible threat of force.

The risk is that a generation's disenchantment with the nation-state and disinterest in our collective defense have resulted in an unquestioned yet massive redirection of resources, both intellectual and financial, to sating the often capricious needs of capitalism's consumer culture. Our loss of cultural ambition, and the diminishing demands we place on the technology sector to produce products of enduring and collective value to the public, have ceded too much control to the whims of the market. As David Graeber, who taught cultural anthropology at Yale and the London School of Economics, observed in an essay published in 2012 in the *Baffler*, "The Internet is a remarkable innovation, but all we are talking about is a super-fast and globally accessible combination of library, post office, and mail-order catalogue." He, and many others, have been left wanting more.

In November 2022, when OpenAI, which has invested billions of dollars into the development of large language models such as ChatGPT, first released its AI interface to the public, the company's policies prohibited the use of its technologies for "military and warfare" purposes, a broad concession to those wary of any entanglements with the soldiers sent into harm's way to defend the nation. After the company changed course in early 2024 and removed the blanket prohibition on military applications, protesters promptly gathered in San Francisco outside the office of Sam Altman, the chief executive officer, with organizers of the protest demanding that OpenAI "end its relationship with the Pentagon and not take any military clients." The engineers building the language models that drive ChatGPT, a spectacular advance in the way computing intelligence approaches problems, are more than content to lend the power of their creation to corporations selling consumer goods yet hesitate when asked to provide more effective software to the U.S. Army and Navy.

The threat of such protest and outrage from the crowd is that it shapes and influences the instincts of leaders and investors across the technology industry, many of whom have been trained to systematically avoid any hint of controversy or disapproval. And the costs of such avoidance—as well as the industry's near-complete capitulation to the whims of the market for direction as to what *ought* to be built, not merely what *can* be built—are significant.

In an essay titled "Big Idea Famine," published in the *Journal of Design and Science* in 2018, Nicholas Negroponte, co-founder of the Massachusetts Institute of Technology's Media Lab, noted the legions of "start-ups today that focus on thoughtless ways to do our laundry, deliver food or entertain ourselves with another app." The challenge, he added, is that "new technologies, real discoveries, and inventions in science and engineering are often trivialized by the start-up process in order to meet the expectations of investors." Many entrepreneurs and armies of extraordinarily talented engineers simply set the hard problems aside. This retreat of ambition has coincided with what the economist Robert J. Gordon has argued has been a significant decline in our rate of productivity as a society in

FIGURE 4

Total Factor Productivity Growth in the United States (1900 to 2014)

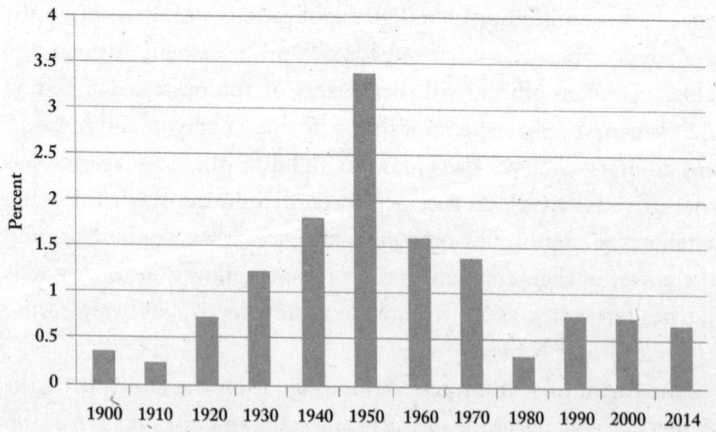

the United States over the past three-quarters of a century. As Gordon has written, in the decades since 1970, technological developments "have mostly occurred in a narrow sphere of activity having to do with entertainment, communications and the collection and processing of information," whereas "for the rest of what humans care about—food, clothing, shelter, transportation, health and working conditions both inside and outside the home—progress has slowed."

There are exceptions to the technology industry's broad retreat of ambition. Elon Musk, for example, has founded two companies, Tesla and SpaceX, among others, that have stepped forward to fill glaring innovation gaps where national governments have stepped back. The challenges of developing a reliable alternative to the internal combustion engine, and to sending rockets into outer space, in another era would have been the comfortable and logical preserve of government. The resources required to confront such challenges are enormous. Yet far too few have been willing to risk their capital or reputations in attempting to address them. The culture almost snickers at Musk's interest in grand narrative, as if billionaires ought to simply stay in their lane of enriching themselves and perhaps providing occasional fodder for celebrity gossip columns. A profile of Musk in the *New Yorker* published in 2023 suggested that the world would be better off with fewer "mega-rich luxury planet builders," decrying his "seeming estrangement from humanity itself."* For years, many were convinced that SpaceX's reusable rockets were "a fool's errand" and that Musk was "flat-out wasting his time," according to a 2015 biography of the founder. Any curiosity or genuine interest in the value of what he has created is essentially dismissed, or perhaps lurks from beneath a thinly veiled scorn. The irony is that many of those who profess most strenuously that they oppose the excesses of capitalism are often the first in line to skewer those who

---

* Musk's critics are often far from the arena, in the words of Theodore Roosevelt—"those cold and timid souls" who know "neither victory nor defeat."

have the audacity to attempt building something that the market has failed to provide. More ambition and seriousness of purpose, not less, are needed. Is the iPhone, for example, our greatest creative if not crowning achievement as a civilization? The object has changed our lives, but it may also now be limiting and constraining our sense of the possible. As Peter Thiel observed in an interview in 2011, the radical and discontinuous leap forward of the Apollo space program, not the incremental advances in the capabilities of consumer gadgets, should be the bar by which we judge ourselves and assess human progress.

A generation of ascendant founders says it actively seeks out risk, but when it comes to public relations and deeper investments in more significant societal challenges, caution often prevails. Why take the chance of entering into the moral morass of geopolitics and courting controversy when you can build another app?

And build apps they did. The proliferation of social media empires across the United States, which systematically monetize and channel the human desire for status and recognition, preying on and programming the young to find rewards in the often fickle affection and approval of their peers, has redirected far too great a share of the efforts and resources of an entire civilization. In 2022, YouTube made $959 million from advertising that was targeted at 31.4 million children under the age of twelve. Instagram made $801 million in a year from that same age-group. We must rise up and rage against this misdirection of our culture and capital. Let us not go gentle into that good night.*

---

* The misdirection of our attention and resources to such endeavors is not the result of some nefarious plot, but rather the consequence of a failure of will and imagination by those at the helm. As a nation, we should move to build, for example, a technological peace corps—an institution through which curious and talented engineering minds whose efforts might otherwise be co-opted to further refine online advertising algorithms could instead be directed to addressing glaring innovation gaps across education, medicine, national defense, and basic science in the United States and abroad.

Our adversaries will proceed with the development of artificial intelligence for the battlefield whether or not we do. The leaders of authoritarian regimes might very well lose their lives if they lose control. Xi Jinping, China's head of state, was born in 1953, four years after the end of the country's communist revolution. At the age of fifteen, he was sent to Liangjiahe, a village to the northeast of Xian in Shaanxi province, where he lived in a cave and was forced to work in the fields, according to an account of his youth. "He ate bitterness like the rest of us," a farmer who knew Xi during those early years told a newspaper in 2012. It was a period of immense social upheaval. Xi's older sister, Heping, might have killed herself in the hands of the Red Guards, the students and others that Mao Zedong at first rallied in support of his revolution and then scrambled to contain in the 1960s. An official government account reveals little, noting only that Heping was "persecuted to death." As a professor of international relations explained in an interview with Evan Osnos of the *New Yorker* in 2022, many of Xi's contemporaries who lived through the Cultural Revolution "concluded that China needed constitutionalism and rule of law, but Xi Jinping said no: You need the Leviathan." The cultivation of hard power, including AI for the battlefield, is a necessity to survive. Xi understands this in a way that those in the West, the self-proclaimed victors of history, often forget.

The American foreign policy establishment has repeatedly miscalculated when dealing with China, Russia, and others, believing that the promise of economic integration alone will be sufficient to undercut their leadership's support at home and diminish their interest in military escalations abroad. The failure of the Davos consensus, the reigning approach to international relations, was to abandon the stick in favor of the carrot alone. Anne Applebaum rightly reminds us that a "natural liberal world order" does not exist, despite our most fervent aspirations, and that "there are no rules without someone to enforce them." Xi and others have wielded and retained power in a way that most of our current political leaders in the West will never understand. Our mistake is to hope that authoritarian regimes, with enough proximity to and encouragement from our own, will realize the error of their ways. But as Henry Kissinger has observed, "The institutions of the West did not spring full-blown from the brow of contemporaries but evolved over centuries."

We must not lose interest in investigating the psychology and worldview of our adversaries, in inhabiting the constraints within which they operate, the risks they face to maintaining control, their personal ambitions, and aspirations for their people. Xi and his family have demonstrated a curiosity and interest in the United States for decades. In 1985, he spent time in Muscatine, Iowa, as part of a delegation from China to the United States, staying in a local family's home. And Xi's only daughter, Xi Mingze, graduated from Harvard in May 2014, using a pseudonym and studying English and psychology. A reporter for a Japanese newspaper said that fewer than ten people were aware of Mingze's real identity while she was at school.

On a visit to the United States in 2015, Xi gave a speech in Seattle in which he recalled reading Henry David Thoreau, Walt Whitman, and Mark Twain when he was young. Ernest Hemingway left a particular impression on him, and Xi remembered *The Old Man and the Sea* with affection. When Xi visited Cuba, he told the audience that he made a trip to Cojímar, a district outside central Havana on the country's northern coast, which had provided inspiration for Hemingway's story of a fisherman and his eighteen-foot marlin. On a later trip, Xi mentioned that he "ordered a mojito," the author's favorite, "with mint leaves and ice." Xi explained that he "just wanted to feel for myself" what Hemingway had been thinking and the place he had been when "he wrote those stories." The leader of a nation with nearly one-fifth of the world's population added that it was "important to make an effort to get a deep understanding of the cultures and civilizations that are different from our own." We would be well advised to do the same.

. . .

The reluctance on the part of the United States and its allies to proceed with the development of more effective and autonomous weapons systems for military use may stem from a justified skepticism of power itself and coercion—a distaste for further investment in the machinery of war by the victors of history. The appeal of pacifism is that it satisfies our instinctive empathy for the powerless. But as Chloé Morin, a French author and former adviser to the country's prime minister, suggested in a recent interview, we might resist the facile urge "to divide the world into dominants and dominated, oppressors and the oppressed." This "moral dualism," in the words of Remi Adekoya, a professor at the University of York in the United Kingdom, leaves many uncomfortable, condemning harm against those who in certain domains occupy positions of power. It would be a mistake, however, and indeed a form of moral condescension, to systematically equate powerlessness with piousness. The subjugated and the subjugators are both equally capable of grievous sin. Yet we still cling to dangerous and pervasive mythologies of a "pacified past," as Lawrence H. Keeley described it in *War Before Civilization*, published in 1996, in which he recounted the history of often brutal violence in preindustrial societies, from the Cheyenne in the Great Plains of North America to the Dani in New Guinea. Keeley, for instance, noted that some indigenous tribes on the American Plains "mutilated their foes' corpses in characteristic ways as a kind of 'signature': the Sioux by cutting throats, the Cheyenne by slashing arms, the Arapaho splitting noses." The Dani in Indonesia, for their part, used mud or grease on their arrowheads to increase the chances of infection for those they shot.

The roots of this moral logic run deep and may be difficult to dislodge. In 1968, Paulo Freire, the Brazilian writer, published *Pedagogy of the Oppressed*, in which he articulated a logic of oppressor and oppressed that continues to structure our intellectual and moral discourse half a century later. One of his central claims was that the oppressed peoples of the world, the underclass, were essentially incapable themselves of violence, or indeed oppression itself. He neutered the dispossessed of moral agency. "Never in history has violence been initiated by the oppressed," he wrote. "It is not the helpless, subject to terror, who initiate terror, but the violent." For him, the subjugated peoples of the world were essentially incapable themselves of victimizing others, only of being victims. But this reductionist insistence on imposing such a totalizing and complete identity on the purportedly powerless may have the unintended consequence of depriving them of moral agency and indeed their humanity as well.

The allure of pacifism, and a potential retreat from deterrence, is that it relieves us of the need to navigate among the difficult and imperfect trade-offs that the world presents. The broader question we face is not whether a new generation of increasingly autonomous weapons incorporating artificial intelligence will be built. It is who will build them and for what purpose. This is the software century, and yet our challenge is that the generation that is most capable and best positioned to construct this next wave of offensive capabilities is also the most content to retreat from projects involving national defense or communal purpose. It is this hollowing out of the American mind—and not only in Silicon Valley, as we will see in the next chapter—that has led us to the current impasse. And it is that hollowing out of the American project that has left us vulnerable and exposed.

Part II

The Hollowing Out of the American Mind

Chapter Five

The Abandonment of Belief

IN 1976, FRANK COLLIN, an ambitious leader in the small but resilient Nazi Party of the United States, planned a march in Skokie, Illinois—an attempt to raise the profile of his organization and build support for his cause. The town, many of whose residents were Jewish and had lived through the war, vehemently opposed the demonstration, and the case went to the courts. The American Civil Liberties Union (ACLU) came to the legal defense of Collin and his fellow Nazis on First Amendment grounds—a move that would be almost unthinkable today. Aryeh Neier, the national executive director of the ACLU at the time, received thousands of letters condemning his organization's decision to defend the free speech rights of Nazis. Neier was born into a Jewish family in Berlin in 1937 and fled from Germany to England along with his parents as a child. He later estimated that thirty thousand ACLU members left the organization as a result of its decision to come to the legal defense of the Nazi demonstrators.

His interest in protecting Collin's right to free speech under the First Amendment was not rooted in an unthinking commitment to liberalism or its values. He instead held two seemingly contradictory yet deeply felt and genuine beliefs—in the abhorrence of Collin's views and in the importance of defending his right to express them against infringement by the state. Neier was interested and willing to stand up for an ideal—something above and beyond his own interests, and one that many would have been content to applaud him for setting aside. "To defend myself, I must restrain power with freedom, even if the temporary beneficiaries are the enemies of freedom," he later wrote. His beliefs had a cost, and their defense required putting the credibility of his organization, and himself, at risk.

A decade before, in September 1963, a similar clash arose in New Haven, Connecticut, where George Wallace, the governor of Alabama and vehement opponent of integration, had been invited to speak by the Yale Political Union, a student organization. Earlier that year, at his inaugural address in January, Wallace had told a crowd in Montgomery, Alabama, that integration must be resisted as a form of "communistic amalgamation," one that would result in a "mongrel unit of one under a single all powerful government." It was in that speech that Wallace said that he would be drawing a "line in the dust," calling for "segregation now, segregation tomorrow, and segregation forever," to cries of support from the crowd.

His potential arrival in New Haven had engulfed the city. Mayor Richard C. Lee decided to send a telegram to Wallace letting him know that he was "officially unwelcome"—an attempt to cancel the event, which many thought would spark violence. Earlier that month, a group of four members of the Ku Klux Klan had used dynamite to bomb the 16th Street Baptist Church in Birmingham, Alabama, killing four girls and injuring nearly two dozen.

Others, however, urged the university not to prevent Wallace from speaking. Pauli Murray, who was pursuing a doctorate at Yale Law School, wrote a letter to Kingman Brewster Jr., the university's president, asking him to permit Wallace to address students on campus. Murray, born in Baltimore in 1910, was a civil rights activist who worked for a period as an attorney at Paul, Weiss, Rifkind, Wharton & Garrison, the New York law firm, and later taught at the Ghana School of Law. She founded, along with Betty Friedan and others, the National Organization for Women in 1966. For Murray, the question of whether Wallace should be permitted to speak on campus was personal. Her father had been committed to the Crownsville State Hospital for the Negro Insane in Maryland, where he was killed in 1922 after "a white guard taunted him with racist epithets, dragged him to the basement, and beat him to death with a baseball bat," according to one account. Murray's maternal grandmother was born into slavery in North Carolina.

Her letter to Brewster was nonetheless direct and brimming with conviction and clarity. She argued that even though she herself had "suffered from the evils of racial segregation," a "possibility of violence is not sufficient reason in law to prevent an individual from exercising his constitutional right." Murray anticipated the risk of allowing for what would later be described as a "heckler's veto" over the speech rights of others—the possibility that debate would be silenced as the result of a fear of the reaction, even a violent one, of a listener. In the modern era, the veto is, of course, wielded with frequency by those who profess offense or discomfort when faced with views other than their own. The Yale Political Union eventually rescinded its invitation to Wallace under pressure from Brewster.

* * * *

Both Neier and Murray, in different contexts and different decades, not only defended the unpopular but risked their reputations, as well as the disapproval of their peers and the public, to stand up for a sort of hard belief, one that was not vulnerable to being abandoned and rationalized away. For Neier and Murray, something more than their own self-preservation and advancement was at stake. Similar tests have presented themselves more recently. But our culture has stepped back from nurturing and encouraging such radical acts of intellectual courage, leaving us with leaders who are increasingly unsure of themselves and unwilling, or perhaps unable, to place much at risk.

In 2023, three university presidents, of Harvard, the University of

Pennsylvania, and the Massachusetts Institute of Technology, were called before Congress in response to protests against Israel's invasion of Gaza following the killing of more than eleven hundred people in Israel and the taking of some 250 hostages. The testimony of the university presidents—two of whom ultimately resigned from their positions—raised issues similar to those that arose in Skokie and New Haven decades ago, including the familiar tension between protecting free speech rights and guarding against attempts to alienate and subjugate the other. Their cautious responses, attempting to preserve space for free speech, captured national and international attention. To many, the presidents were far too tepid in articulating their opposition to overtly hostile calls and intimidation of students on campus. As Maureen Dowd pointed out in the *New York Times*, Elizabeth Magill, president of the University of Pennsylvania, "offered a chilling bit of legalese" when she was asked whether calls for the genocide of Jews constituted harassment. Magill responded, "It is a context-dependent decision."

The presidents were wholly unaware of the internal contradictions of their position—contradictions stemming from their commitment to free speech, on the one hand, but also the eagerness of their institutions in various other contexts to carefully patrol the use of language for fear of causing offense. Their halting testimony was marked by cool precision and calculation—embodying the archetype of the new administrative class, clinical and careful and above all without feeling.

The testimony exposed a fundamental challenge that we, in the United States and the West, face. A broad swath of leaders, from academic administrators and politicians to executives in Silicon Valley, have for years often been punished mercilessly for publicly mustering anything approaching an authentic belief. The public arena—and the shallow and petty assaults against those who dare to do something other than enrich themselves—has become so unforgiving that the republic is left with a significant roster of ineffectual, empty vessels whose ambition one would forgive if there were any genuine belief structure lurking within.

The unrelenting scrutiny to which contemporary public figures are now subjected has also had the counterproductive effect of dramatically reducing the ranks of individuals interested in venturing into politics and adjacent domains. Advocates of our current system of ruthless exposure of the private lives of often marginally public figures make the case that transparency, one of those words that has nearly become meaningless from overuse, is our best defense against the abuse of power. But few seem interested in the very real and often perverse incentives, and disincentives, we have constructed for those engaging in public life.

The stifling regime of disclosure and punishment for authentic intellectual risk-taking that we impose on would-be leaders leaves little room for capable and original thinkers whose principal motivation is something other than self-promotion, and who often lack a willingness to subject themselves to the theater and vicissitudes of the modern public sphere. It is "the proliferation of frenzies and expansion of the range of personal issues subject to scrutiny," as one political scientist who has attempted to measure the fall in quality of political candidates as a result of increasingly invasive media coverage of public figures has put it, that "raises the expected cost to good people of running for public office." In 1991, Larry Sabato, a professor of politics at the University of Virginia, joked that we were not far from the moment at which the press would pounce on a candidate "for using an express checkout lane when purchasing more than the ten-item limit."

The expectations of disclosure have increased steadily for more than half a century and have brought essential information to the voting public. They have also contorted our relationship with our elected officials and other leaders, requiring an intimacy that is not always related to assessing their ability to deliver outcomes. Americans, in particular, "have overmoralized public office," as an editorial in *Time* magazine warned decades ago in 1969, and "tend to equate public greatness with private goodness." The risk is that the political realm, and the empowerment that one can feel by participating in the democratic process, becomes more about our own psychological need for self-expression than actual governance. Those who look to the political arena to nourish their soul and sense of self, who rely too heavily on their internal life finding expression in people they may never meet, will be left disappointed. We think we want and need to *know* our leaders. But what about results? The likability of our elected leaders is essentially a modern preoccupation and has become a national obsession, yet at what cost?

In 1952, Richard Nixon, who was then General Dwight Eisenhower's vice presidential running mate, gave what would become known as the Checkers speech, after his black-and-white cocker spaniel, disclosing to the American public that he owned a home in Whittier, California, at a cost of $13,000, on which he had an outstanding mortgage of $3,000. He had been accused of improperly using political funds for personal benefit and had felt the need to try to clear the air. In that moment, the country was for the first time introduced to a new and striking level of granularity in the disclosures that it required from its politicians, and perhaps the beginning of a decline in the quality of those willing to come forward and submit to the spectacle. His wife reportedly asked Nixon, affecting a certain naïveté, faux or otherwise, "Why do you have to tell people how little we have and how much we owe?" Her husband replied that politicians were destined to "live in a goldfish bowl." But the systematic elimination of private spaces, even for our public figures, has consequences, and ultimately further incentivizes only those given to theatrics, and who crave a stage, to run for office. The candidates who remain willing to subject themselves to the glare of public service are, of course, often interested more in the power of the platform, with its celebrity and potential to be monetized in other ways, than the actual work of government.

THE ABANDONMENT OF BELIEF

* * *

The current system of disclosure and scrutiny to which we subject our leaders is not limited to university presidents or elected officials. It has also permeated the ranks of Silicon Valley and the corporate world. An entire generation of executives and entrepreneurs that came of age in recent decades was essentially robbed of an opportunity to form actual views about the world—both descriptive, what it is, and normative, what it should be—leaving us with a managerial class whose principal purpose often seems to be little more than ensuring its own survival and re-creation.

The atrophying of the mind, and the self-editing that often accompanies such decay, are corrosive to real thought. The result is that corporations selling consumer goods feel the need to develop and indeed broadcast their views on issues affecting our moral or interior lives, while most software companies with the capacity and, perhaps, duty to shape our geopolitics remain conspicuously silent.*

Palantir builds software and artificial intelligence capabilities for defense and intelligence agencies in the United States and its allies across Europe and around the world. Our work has been controversial, and not everyone will agree with our decision to build products that enable offensive weapons systems. But we have made a choice, notwithstanding its costs and complications.

By contrast, the congressional testimony by the university presidents exposed the bargain that contemporary elite culture has made

---

* Appeals to virtue and character, having been excluded for the most part from the civic and political realms, have migrated, or rather, been co-opted and appropriated, by the corporate. In 2013, Ram Trucks produced a television commercial featuring a speech titled "So God Made a Farmer" from 1978 by Paul Harvey, a radio broadcaster born in Tulsa, Oklahoma. The speech hailed the American farmer, who, among other things, was "willing to sit up all night with a newborn colt, and watch it die, then dry his eyes, and say maybe next year." It was poignant and powerful— yet all in service of selling a pickup truck. We have, quite unwittingly, ceded direction over our interior lives, the development of our moral selves, to the market.

to retain power—that belief itself, in anything other than oneself perhaps, is dangerous and to be avoided. The Silicon Valley establishment has grown so suspicious and fearful of an entire category of thought, including contemplations on culture or national identity, that anything approaching a worldview is seen as a liability. The shallow and thinly veiled nihilism of a corporate slogan such as "don't be evil," which Google adopted when the company went public in 2004 and later exchanged for the similarly banal "do the right thing," captures the views of a generation of extraordinarily talented software engineers who were taught to prize the identification of and resistance to evil over the more difficult and often messy task of navigating the world in all of its imperfection. As the French author Pascal Bruckner has written, when we lack "the power to do anything, sensitivity becomes our main aim," and thus "the aim is not so much to do anything, as to be judged."

The problem is that those who say nothing wrong often say nothing much at all. An overly timid engagement with the debates of our time will rob one of the ferocity of feeling that is necessary to move the world. "If you do not feel it, you will not get it by hunting for it," Goethe reminds us in *Faust*. "You will never touch the hearts of others, if it does not emerge from your own."

Our culture has for the most part successfully pounded down any notes or errant hints of zeal and feeling in many of those leading our most significant institutions. And what remains beneath the polish is often unclear. We later learned that WilmerHale, one of the nation's most respected law firms, prepared and advised both Claudine Gay of Harvard and Elizabeth Magill of the University of Pennsylvania for their testimony before Congress. And both of them lost their jobs. The clinical approach of the presidents, and their trust in legal specialists to guide them in what would essentially become a referendum on their convictions, are reminders of the perils of delegating the waging of political battle to legal referees at its margins. Others suggested that the questioning and treatment of the college presidents was unfair. It might have been. But as Lawrence Summers, the former president of Harvard, correctly pointed out, even if we acknowledge that the congressional inquisition of the college presidents was a form of "performance art," we should expect more from our leaders on that significant of a stage.

When we require the systematic elimination of the thorns, barbs, and flaws that necessarily accompany genuine human contact and confrontation with the world, we lose something else. The work of Erving Goffman, a Canadian-born sociologist, on what he described as "total institutions," is instructive here. In a collection of essays published in 1961 titled *Asylums*, Goffman defined such institutions, which include prisons and mental hospitals, as places "where a large number of like-situated individuals, cut off from the wider society for an appreciable period of time, together lead an enclosed, formally administered round of life." The same might be said of some of our nation's most elite universities, which have nominally and belatedly opened their doors to a far broader swath of participants but whose internal cultures remain remarkably cloistered and walled off from the world.

In the late 1960s, an earlier generation of university administrators, including Kingman Brewster Jr. at Yale, took a different path when it came to confronting and embracing a challenge to entrenched power and elite privilege. A series of civil rights demonstrations involving the Black Panthers and others engulfed Yale's campus in May 1970, and at least two bombs exploded in the school's ice hockey rink. There was a willingness, however, by Brewster and others to venture into the ethical morass of the moment in a way that would ensure a swift and summary cancellation in the United States today. In April 1970, at a meeting of hundreds of Yale faculty members in New Haven, Connecticut, Brewster said that he was "skeptical of the ability of black revolutionaries to achieve a fair trial anywhere in the United States," according to a report in the *Times* the following day. He had ventured into the conflagration, not away from it. Spiro Agnew, the vice president of the United States, promptly called for Brewster to resign. He did not, however, and Brewster not only kept his job but ultimately emerged stronger. As Ralph Waldo Emerson once said, "When you strike at a king, you must kill him."

Allan Bloom, who taught at the University of Chicago, more than three decades ago articulated the challenge that we currently face in his 1987 polemic, *The Closing of the American Mind*. Our commitment to "openness," a vital and uncontroversially necessary good, he wrote, "has driven out the local deities, leaving only the speechless, meaningless country." Bloom continued: "There is no immediate, sensual experience of the nation's meaning or its project, which would provide the basis for adult reflection on regimes and statesmanship. Students now arrive at the university ignorant and cynical about our political heritage, lacking the wherewithal to be either inspired by it or seriously critical of it." In the late 1980s, Bloom was focused on the interior and intellectual lives of university students. It is now those students who are our administrators. And the culture in which those administrators have been raised has been unforgiving, systematically punishing anything approaching moral courage and incentivizing its opposite. In this way, the university presidents are victims of their and our collective focus on the policing of language and by extension thought, combined with the enforcement of elaborate yet unpublished codes regarding speech and behavior—that together deprive individuals of the habit and instinct required to develop sincerely held and authentic beliefs, as well as the gall to express them.

Perry Link, the former professor of East Asian studies at Princeton whose work in the 1990s was vital in exposing the massacre at Tiananmen Square in Beijing, has noted that the Soviet leadership went to great lengths to document and detail the proscriptions of the day, even publishing "periodic handbooks that listed which specific phrases were out of bounds."* The means by which the Chinese government patrolled the boundaries of speech, however, were far more subversive in Link's view, and in many ways more closely approximate the contemporary model of attempts to constrain speech in the United States. Link wrote that the Chinese government "rejected these more mechanical methods" of censorship used by the Soviet regime "in favor of an essentially psychological control system," in which each individual must assess the risk of a statement against what Link describes as "a dull, well-entrenched leeriness" of disapproval by the state.

Amid the campus protests across the United States in 2024 following Israel's invasion and bombardment of Gaza, a growing number of student protesters began concealing their faces with scarves and masks. Their rationale was that exposure of their identities would jeopardize their futures, from depriving them of job opportunities to facing criticism on social media. A student protester at Northwestern, in Evanston, Illinois, told a reporter in May 2024 that the potential costs were too great to risk being identified. "If I give my name, I lose my future," he said. But is a belief that has no cost really a belief? The protective veil of anonymity may instead be robbing this generation of an opportunity to develop an instinct for real ownership over an idea, of the rewards of victory in the public square as well as the costs of defeat.

Michael Sandel, a professor at Harvard, anticipated the contradictions that arise from our fierce commitment in the West to classical liberalism, and its elevation if not preference for individual rights at the expense of anything approaching collective purpose or

---

* One Soviet directive from the 1920s listed ninety-six categories of information that were prohibited, including facts and statistics regarding "sanitary conditions in places of incarceration," "clashes between the authorities and peasants during the implementation of tax and fiscal measures," as well as "cases of mental derangement caused by unemployment and hunger."

identity, as well as our cultural reluctance to venture into many of the most meaningful and significant moral debates of our time. It is this fundamental abdication of responsibility for articulating a coherent and rich vision of the world, and of shared purpose—the systematic dismantling of the West—that has left us unable to confront issues with moral clarity or true conviction. And the consequences of this inability or unwillingness to enter into such debates, "where liberals fear to tread," as Sandel famously put it, are now being made clear. "Where political discourse lacks moral resonance, the yearning for a public life of larger meanings finds undesirable expressions," he wrote in *Liberalism and the Limits of Justice*. As a result, our broader cultural discourse shrinks down into something small and petty, becoming "increasingly preoccupied with the scandalous, the sensational, and the confessional," Sandel added. His broader critique was that a certain narrowness of modern liberalism "is too spare to contain the moral energies of a vital democratic life," and "thus creates a moral void that opens the way for the intolerant" and "the trivial." That void, haunting and fearsome, is now being revealed.

Chapter Six

Technological Agnostics

THE CURRENT LEADERS OF Silicon Valley, who have constructed the technical empires that now structure our lives, were for the most part raised in a culture nominally reverent of the requirements of justice. But discussion of the vast realm of questions that afflict our moral lives beyond adherence to the basics—a commitment to equality, of some sort, and certainly the rights of others—was essentially off limits. Any inquiry into what constituted a good or virtuous life, of what allegiance, to one's country, for example, meant in the modern era, was beyond the meadow of permissible discourse. This generation, the first significant set of graduates from a far more open university system in the United States, was reluctant to limit its options, to exclude the views of others, and to stake out ideological and political stands. The pursuit of optionality, both in their business and in their intellectual lives, if not their personal and romantic choices as well, was paramount. The principal affiliation of this generation of builders was to the businesses they themselves were building. And at school, the subtext of their education from an early age had been that an overly fervent reverence for the American project, let alone the West, should be viewed with skepticism.

Amy Gutmann, who taught at Princeton through the 1980s and 1990s, captured the logic of the era when she argued that "our primary moral allegiance is to no community," national or otherwise, but rather "to justice" itself. The ideal at the time, and still for many today, was for a sort of disembodied morality, one unshackled from the inconvenient particularities of actual life. But this move toward the ethereal, the post-national, and the essentially academic has strained the moral capacity of our species. These cosmopolitan and technological elites in the developed world were citizens of no country; their wealth and capacity for innovation had, in their minds, set them free. As Manuel Castells Oliván, a Spanish sociologist, has written, "Elites are cosmopolitan, people are local." The instinct of this generation of technology founders and programmers was to avoid forgoing paths, taking sides, alienating anybody. This cult of optionality, however, has been crippling, constraining the development of young minds and condemning them to a sort of perpetual preparation for a battle they may never fight. The future belongs to those who scuttle the ships.* The ubiquitous off-ramps and backup plans among the current generation, and instinct to burnish the rough edges off of one's opinions, stand in opposition to throwing oneself into an endeavor with the abandon, nearly reckless, that is required to succeed, or at least fail in a sufficiently substantial way that provokes development.

The current emerging technological class in the United States—the masters of this new universe that we inhabit, willingly or otherwise—often points to software and artificial intelligence as our salvation. They believe, to be sure, but principally in themselves and in the power of their creations, stopping short of entering into a discourse with the most significant questions of our time, including

---

* Hernán Cortés, the Spanish governor of Cuba in the sixteenth century, did not in fact burn his ships as is often suggested but rather likely had his crew run them aground in 1519 on the beach of Veracruz, on the eastern coast of Mexico. He destroyed at least nine of his ships in an attempt to deprive his men of the opportunity to mutiny and sail back to Cuba on their own, providing them with the option to return home on a lone remaining vessel but only, according to one historian, "to discover who the cowards and untrustworthy ones were."

the broader project of the nation and its reason for being. They are building, but we should ask for what purpose and why. President Eisenhower warned in his farewell address in January 1961 of both the rise of a "military-industrial complex" and the "danger that public policy could itself become the captive of a scientific-technological elite." Our current era of innovation has been dominated by the indiscriminate construction of technology by software engineers who are building simply because they can, untethered from a more fundamental purpose.

There is a purity to this desire to construct for the sake of construction. And the amount of sheer creative production is impossible to deny. Mark Zuckerberg, who co-founded Facebook, now Meta, in 2004, demonstrated to the world a level of scaling—from literally dozens to hundreds to thousands to millions to billions of users—that humanity hardly understood to be possible and is still difficult to comprehend. His platform has repeatedly broken through purported ceilings in its potential, confounding supporters and critics alike. After *The Social Network* was released in 2010, Zuckerberg took issue with the film's attempt to frame his interest in building what would become Facebook as a desire for status or even the affections of the opposite sex. "They just can't wrap their head around the idea that someone might build something because they like building things," he said at a talk at Stanford University in October 2010. He captured the views of a generation of software engineers and founders, whose principal and animating interest was the act of creation itself—decoupled from any grand worldview or political project. These were the technological agnostics.

Our educational institutions and broader culture have enabled a new class of leaders who are not merely neutral, or agnostic, but whose capacity for forming their own authentic beliefs about the world has been severely diminished. And that absence leaves them vulnerable to becoming instruments for the plans and designs of others. An entire generation is at risk of being deprived of the opportunity to think critically about the world or its place in it. It is this *productization* of the American mind, in addition to its closing, that we must guard against. A significant subset of Silicon Valley today undoubtedly scorns the masses for their attachment to guns and religion, but that subset clings to something else—a thin and meager secular ideology that masquerades as thought.

It may be axiomatic in contemporary culture that all views should be tolerated, but we need to admit that even the faintest whiff of actual religion in certain circles, unironic belief in something greater—in many corporate boardrooms and certainly the halls of our most selective colleges and universities—is looked down upon as essentially preindustrial and retrograde. This shift has been happening for decades. The elite's intolerance of religious belief is perhaps one of the most telling signs that its political project constitutes a less open intellectual movement than many within it would claim. As Stephen L. Carter, a professor at Yale Law School, wrote in his book *The Culture of Disbelief*, published in 1993, from the perspective of the educated ruling class in this country, "taking religion seriously is something that only those wild-eyed zealots do." Carter noted that the roots of the contemporary skepticism of religion are essentially modern, beginning with Freud perhaps, who viewed religion as a sort of obsessive impulse. In an essay titled "Obsessive Actions and Religious Practices," published in 1907, Freud wrote that the "formation of a religion," with its oscillating focus on guilt and atonement from sin, itself "seems to be based on the suppression, the renunciation, of certain instinctual impulses." It is perhaps that same hostility, often flagrant, to religion in elite culture that holds back the development of belief in the current generation.

There is no question that an unwillingness to revise one's views in light of new evidence is itself an impediment to progress. As the German physicist Max Planck said, "A new scientific truth does not triumph by convincing its opponents and making them see the light, but rather because its opponents finally die." The miracle of the West is its unrelenting faith in science. That faith, however, has crowded out something equally important, the encouragement of intellectual courage, which sometimes requires the fostering of belief or conviction in the absence of evidence.

We have grown too eager to banish any sentiment or expression of values from the public square. The educated class in the United States was content to abstain from engagement with the content of the American national project: What is this nation? What are our values? And for what do we stand? This great secularization of postwar America was cheered by many on the left, either privately or publicly, who saw the systematic eradication of religion from public life as a victory for inclusion. And a victory, in that narrow sense, it was. But the unintended consequence of this assault on religion was the eradication of any space for belief at all—any room for the expression of values or normative ideas about who we were, or should become, as a nation. The soul of the country was at stake, having been abandoned in the name of inclusivity. The problem is that tolerance of everything often constitutes belief in nothing.

We unwittingly deprived ourselves of the opportunity to critique any aspects of culture, because all cultures, and by extension all cultural values, were sacred. After decades of debate, the postmodernist impulse has run its course and exposed its limits. As Fukuyama has written, "If all beliefs are equally true or historically contingent, if the belief in reason is simply an ethnocentric Western prejudice, then there is no superior moral position from which to judge even the most abhorrent practices—as well as, of course, no epistemological basis for postmodernism itself." The postwar move to stamp out belief in America was an overcorrection and left us vulnerable as a society. Was America nothing more than a vehicle through which a newly globalized and educated elite could enrich itself?

Amid the ongoing assault on belief, many Americans have remained essentially ambivalent about the move—not because they are zealots or harbor secret prejudices. They are rather rightly wary and skeptical of the constraints that had been placed on their ability to speak affirmatively on any number of issues, given that speech and language were now being patrolled by bands of secular warriors for any potential violations, however slight, of the new prime directive—which was to offend no one and, consequently, to tread cautiously whenever advocating for a view that might privilege one way of life, one set of values, over another. As a formal matter, dissent was still tolerated. But such tolerance was fickle, and indeed shallow and thin.

• •• •

The employees at Google who resisted leveraging the machinery of their company in service of building software for the U.S. military know what they oppose but not what they are for. The problem that we are describing is not a principled commitment to pacifism or nonviolence. It is a more fundamental abandonment of belief in anything. The company, at its core, builds elaborate and extraordinarily lucrative mechanisms to monetize the placement of advertising for consumer goods and services that accompany search results. The service is vital and has remade the world. But the business, and a significant subset of its employees, stop short of engaging with more essential questions of national purpose and identity—with an affirmative vision of what we want to and should be building as part of a national project, not simply an articulation of the lines that one will not cross. They remain content to monetize our search histories even as they decline to defend our collective security.

Google, of course, along with any number of Silicon Valley's largest technology enterprises, owes its existence in significant part to the educational culture, as well as the legal protections and capital markets, of the United States. The personal computer itself, as well as the internet, was the result of military funding and support in the 1960s from the Defense Advanced Research Projects Agency, a division of the U.S. Department of Defense. In her book *The Entrepreneurial State*,

Mariana Mazzucato, an economics professor at University College London, calls out this collective amnesia in the Valley, noting that the U.S. military's role has "been forgotten" by this era's software titans, who have rewritten history in order to place themselves at its center and exclude and diminish the role of government in fostering and sustaining innovation. And in the absence of any larger project for which to fight, many simply turned elsewhere, not out of some moral failing, but because of the transformation of our most hallowed educational institutions into administrative caretakers, not vessels of culture.

Our reluctance to take on the larger questions has left an enormous amount of talent and zeal on the sidelines. Entire swaths of our generation's greatest minds have drifted, some more willingly than others, into a narrow subset of industries. A survey conducted in 2023 of graduating seniors at Harvard University, for instance, found that nearly half of the entire class was headed for jobs in finance and consulting. Only 6 percent of graduates of Harvard College in 1971 went into those two professions after graduation,

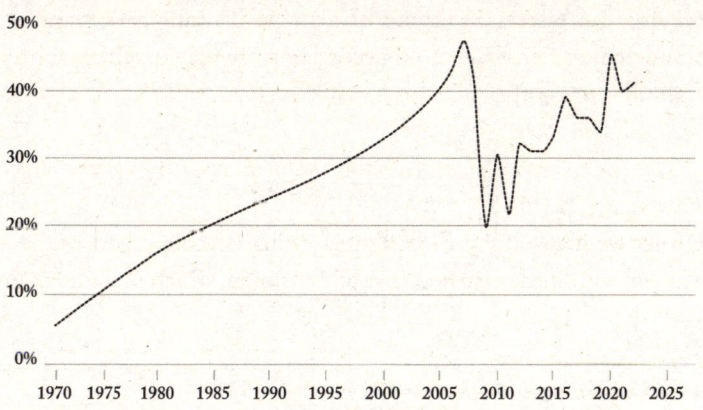

FIGURE 5
Percentage of Harvard Graduates Bound for
Finance or Consulting (1971 to 2022)

according to an analysis by the *Harvard Crimson*. That proportion rose steadily in the 1970s and 1980s, peaking at 47 percent in 2007 just before the financial crisis.

The instrumentalization of American higher education continues unchecked. The number of graduating college seniors who earned a degree in the humanities fell from 14 percent in 1966 to 7 percent in 2010. At the same time, enrollment in computer science and engineering majors has been rising steadily over the past decade, with 51,696 students majoring in those fields in 2014 and 112,720 students in 2023, more than doubling. We need engineers who are engaged with and curious about the world, the movement of history and its contradictions, not merely skilled at programming.

The market has spoken, we tell ourselves, essentially abdicating responsibility for this massive shift in the ambitions and direction of a generation of capable and well-meaning minds. Some graduates, of course, are convinced that they are involved in a broader project. But the mere association of oneself with an ideology or political movement—and resulting feeling of adjacency to engagement and proximity to action—too often masquerade as actual belief or thought. Results need to matter. As Henry Kissinger reminded us, nations "should be judged on what they did, not on their domestic ideology."* The systematic expression and investigation of one's own beliefs—the essential purpose of genuine education—remain our best defense against the mind becoming a product or vehicle for the ambitions of another.

* * *

Earlier we invoked the F-35 fighter jet manufactured by Lockheed Martin, with its anticipated cost of $2 trillion, which includes com-

---

* The cultures and institutions of the world should indeed be judged "by their fruits"—the product and output of their labor. Matt. 7:16.

ponents, from engines to wings, that are manufactured in nearly every one of the fifty U.S. states. The airplanes are made from 300,000 individual parts that are produced by more than eleven hundred suppliers. The parts include $100,000 titanium and aluminum panels that cover the outside of the fuselage made in Phoenix, an $11 million engine made by Pratt & Whitney in East Hartford, Connecticut, and a $300,000 air compressor from a company in Fort Wayne that enables the release of bombs. The breadth and distribution of that supply chain, and its economic benefits, are part of the reason Congress has continued to vote in favor of the program's extension and funding. But what will happen when the defense products of the future, including the artificial intelligence software that will enable the battles of this century, are made by an increasingly concentrated set of companies in Silicon Valley—a sliver of land in a single part of the country? How will the state ensure that this engineering elite remains subservient and accountable to the public?

The fifty most valuable technology companies in the world were worth a combined $24.8 trillion as of 2024. American firms accounted for 86 percent of that total value, or $21.4 trillion. In other words, the United States is responsible for generating nearly nine out of every ten dollars in value of the world's top technology companies. And of those fifty firms, nearly all of the most valuable ones—including Apple ($3.5 trillion), Microsoft ($3.2 trillion), Nvidia ($3.0 trillion), Alphabet ($2.1 trillion), Amazon ($2.0 trillion), Meta ($1.4 trillion), and Tesla ($0.8 trillion)—have roots either in Silicon Valley or on the West Coast. And that level of concentration of wealth and influence—a level that has never before been seen in modern economic history—is only set to increase.[*]

We have made the mistake of allowing a technocratic ruling class

---

[*] The value of the American technology sector, as measured by the market capitalization of all U.S. tech companies, surpassed that of the entire European market in August 2020, according to a survey by an investment bank at the time.

FIGURE 6

The Very Long Term: Estimated GDP Per Capita Worldwide (AD 1 to 2003)

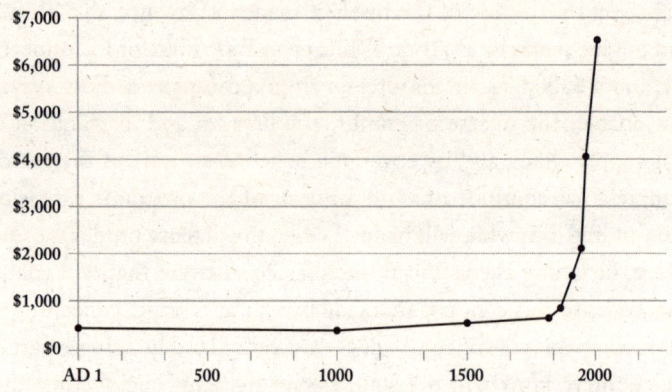

to form and take hold in this country without asking for anything quite substantial in return. What should the public demand for abandoning the threat of revolt? The engineers and entrepreneurs of the Valley have been permitted vast license over broad swaths of the economy, but what should the public ask for in exchange? Free email is not enough.

The broader risk for any country is that elite power structures harden and calcify. In *The Protestant Establishment*, published in 1964, the sociologist E. Digby Baltzell articulated an argument that might feel uncomfortably close to that of many in this country's ruling class today.* In Baltzell's view, an aristocracy driven by talent is

---

* Those who bristle at descriptions of a coastal or transatlantic elite ought to consider how far we have traveled as a nation since 1937, when Ferdinand Lundberg published *America's 60 Families*, which made the case that the United States was "owned and dominated" by a first tier of "its sixty richest families," including the Astors, Du Ponts, Mellons, and Vanderbilts, followed by a second, ancillary tier of "ninety families of lesser wealth."

an essential feature of any republic. The challenge is ensuring that such aristocracies remain open to new members and do not descend into mere caste structures, which close their ranks along racial or religious lines. "If an upper class degenerates into a caste," Baltzell wrote, "the traditional authority of an establishment is in grave danger of disintegrating, while society becomes a field for careerists seeking success and affluence." The challenge for any organization, and indeed nation, is finding ways of empowering a group of leaders without incentivizing them to spend more effort guarding the trappings and perquisites of office than advancing the goals of the group. The caste structures that have formed within countless organizations around the world—from federal bureaucracies to international agencies to academic institutions and Silicon Valley technology giants—must be challenged and dismantled if those institutions have any hope of survival over the long term.

In the end, the nation, this collective attempt at not merely self-governance but the construction of a shared and common life if not purpose, will decide whether it wants Silicon Valley to believe in anything other than the power of its own creations. The technology companies that this country has built have for the most part deftly navigated around any issues that would draw undue scrutiny or unwanted attention; the hallmark of their mode of being is avoidance and, often, silence.

The current silence is a symptom of a broader reluctance to offend and to permit ourselves and those around us to err. In one particularly haunting scene from George Orwell's *1984*, Winston Smith, his protagonist, finds himself wandering through a wooded area, seemingly far from the reach of the state's dystopian minders. Even then, secluded and almost assuredly free from observation, Smith imagines that a microphone might be concealed in the trees, through which "some small, beetle-like man" would be "listening intently." The scene is only nearly fiction. In East Germany, the state security service, known as the Stasi, was rumored to have placed microphones in the trees over ping-pong tables in Berlin's parks, to catch snippets of conversations.

The dystopian future that Orwell and others have imagined may be near, but not because of the surveillance state or contraptions built by Silicon Valley giants that rob us of our privacy or most intimate moments alone. It is we, not our technical creations, who are to blame for failing to encourage and enable the radical act of belief in something above and beyond, and external to, the self. The speed and enthusiasm with which the culture skewers anyone for their perceived transgressions and errors—with which we descend on one another for deviations from the norm—only further diminishes our capacity to move toward truth.

The reluctance of several generations of educators, in particular, but also our political and business leaders, to venture into a discussion about the good, as opposed to merely the right, has left a gap that risks being filled by others, demagogues from both the left and the right.* Such reluctance was born of a desire to accommodate all views and values. But a tolerance of everything has the tendency to devolve into support of nothing. The antiseptic nature of modern discourse, dominated by an unwavering commitment to justice but deeply wary when it comes to substantive positions on the good life, is a product of our own reluctance, and indeed fear, to offend, to alienate, and to risk the disapproval of the crowd. Yet there is too much that lies "beyond justice," in the words of Ágnes Heller, the Hungarian philosopher born in Budapest in 1929. As Heller writes, "Justice is the skeleton: the good life is the flesh and blood." The implications for everything from technology to art are significant.

---

* See John Rawls, "The Priority of Right and Ideas of the Good," *Philosophy & Public Affairs* 17, no. 4 (Autumn 1988): 252, 256 (for a discussion of "the right," which concerns the most fundamental requirements of justice, as opposed to "the good," that is, the many and divergent "views of the meaning, value, and purpose of human life").

We have withdrawn just as much from making ethical judgments about the good life as we have aesthetic judgments about beauty. The postmodern disinclination to make normative claims and value judgments has begun to erode our collective ability to make descriptive claims about truth as well. In *The Twilight of American Culture*, Morris Berman acknowledged that "the deconstructionists were right," in the sense that the context in which a text is written certainly matters, as does its author, and that much of what had passed for *objective* inquiry in academia and elsewhere had been just the opposite. "The problem arises when this position is pushed to the limit," he wrote, "such that you abandon the search for truth and even deny it exists, repudiate the reality of history and intellectual tradition." Our present unwillingness to pronounce, to have a view, and to venture toward the flame, not away from it, risks leaving us adrift.

In a different era, and when confronted with a different sort of test, the American public—enraptured as it was with the prosecutorial zeal and proselytizing of Joseph R. McCarthy, the junior senator from Wisconsin—ultimately came to the conclusion that its purported shepherd was corrupt. We must again look inward, not to our political leaders, many of whom have been complicit in our present descent, but to us, the public itself, for failing to rise up, for failing to resist the hollowing out of our American mind. On March 9, 1954, Edward R. Murrow, the legendary CBS broadcaster of the age, delivered his blistering critique of Senator McCarthy, helping close the chapter on the crusader's particularly enthralling and virulent form of persecution. As Murrow reminded us, quoting Shakespeare's *Julius Caesar*, "The fault, dear Brutus, is not in our stars but in ourselves."

The challenge today will again require a public reckoning with the wisdom of continuing an intellectual war on the concept of the nation, and perhaps nationality itself, that was begun a century ago and whose effects can still be seen today. What began as a noble search for a more inclusive conception of national identity and belonging—and a bid to render the concept of "the West" open to any entrants interested in advancing its ideals—over time expanded into a more far-reaching rejection of collective identity itself. And that rejection of any broader political project, or sense of the community to which one must belong in order to accomplish anything substantial, is what now risks leaving us rudderless and without direction.

Chapter Seven

A Balloon Cut Loose

In December 1976, at a meeting of the American Historical Association in Washington, D.C., Fredric L. Cheyette, a professor of medieval European history at Amherst College, delivered an address calling for the abandonment of the canonical courses on Western civilization that had once been a required rite of passage for undergraduates in American higher education. The debate regarding the survey courses, affectionately and often otherwise known as Western Civ, had been gathering momentum on college campuses for decades, particularly after the end of the war in the 1950s and 1960s.

The question was what, if anything, undergraduates at the country's colleges and universities should learn about Western civilization—about ancient Rome and Greece, through the emergence of the modern form of the nation-state in Europe, and onto our own experiment in the new republic of America. More fundamentally, the issue was whether the concept itself of Western civilization was coherent and substantial enough to hold real meaning in the educational context. The courses spawned an entire subculture of debate about their role and place on campus for nearly half a century, a debate which would become a harbinger of the cultural divide that continues to reveal itself today. And the history of their demise, lost to many in the Valley, suggests the roots of our current predicament.

The issue was not merely what college students ought to be taught, but rather what the purpose of their education was, beyond merely enriching those fortunate enough to attend the right school. What were the values of our society, beyond tolerance and a respect for the rights of others? What role did higher education have, if any, in articulating a collective sense of identity that was capable of serving as the foundation for a broader sense of cohesion and shared purpose? The generations that would go on to build Silicon Valley, to spur the computing revolution, came of age during what would become a massive reassessment of the value of the nation and indeed the West itself.

The traditionalists argued that undergraduates required some basic exposure to thinkers and writers such as Plato and John Stuart Mill, if not also Dante and Marx, in order to understand the freedoms that those students themselves enjoyed and the place in the world that they inhabited. The urge by many at the time to construct a coherent narrative from an enormously fractured historical and cultural record was immense. The supporters of a core curriculum in the Western tradition argued somewhat pragmatically that the American republic required the construction of a shared patrimony or sense of American identity among a cultural elite that was increasingly drawing from a more diverse swath of the population. William McNeill, for example, a historian who began teaching at the University of Chicago in 1947, argued that the construction of a unified canon of texts and narratives, if not mythologies, gave students "a sense of common citizenship and participation in a community of reason, a belief in careers open to talent, and a faith in a truth susceptible to enlargement and improvement generation after generation." The virtue of a core curriculum situated around the Western tradition was that it facilitated and indeed made possible the construction of a national identity in the United States from a fractured and disparate set of cultural experiences—a form of civic religion, tethered largely to truth and history across the centuries but also aspirational in its desire to provide coherence to and grounding of a national endeavor.

Those opposed to the aging survey courses, including Cheyette at Amherst, argued against what they believed was an essentially fictitious grand narrative regarding the arc and development of Western civilization, making the case that such a curriculum was too exclusionary and incomplete to impose on students. Kwame Anthony Appiah, a professor of philosophy at New York University and critic of the entire conception of "the West," would later argue that "we forged a grand narrative about Athenian democracy, the Magna Carta, Copernican revolution, and so on," building to the crescendo of a conclusion, notwithstanding evidence to the contrary, that "Western culture was, at its core, individualistic and democratic and liberty-minded and tolerant and progressive and rational and scientific." For Appiah and many others, the idealized form of the West was a story, riveting perhaps and compelling at times, but a narrative nonetheless, and one that had been imposed, and awkwardly foisted and fitted, onto the historical record, rather than emerging from it.

It was also, of course, very much in dispute where "the West" was even located, that is, which countries counted. When Samuel Huntington published his essay "The Clash of Civilizations?" in *Foreign Affairs* in 1993, he included a map of Europe with a line that William Wallace, then a research fellow at Oxford University, had argued showed the extent of Western Christianity's advance as of 1500.

Most scholars resisted what they described as Huntington's facile division of the world into seven, or possibly eight, discrete "civilizations."* But while his frame was certainly reductionist—indeed, its appeal stemmed from its apparent precision—the wholesale revolt against Huntington would end up crowding out most

---

* Huntington's count of "major civilizations" included "Western, Confucian, Japanese, Islamic, Hindu, Slavic-Orthodox, Latin American and possibly African."

FIGURE 7
The Huntington-Wallace Line

serious normative discussions about the role of culture in shaping everything from international relations to economic development. Where were the fault lines between cultures? Which cultures were aligned with the advancement of the interests of their publics? And what should be the role of the nation in articulating or defending a sense of national culture? The entire terrain would become verboten to scholars who had thoughts of tenure.

. . .

By the late 1970s, the traditionalists had lost the battle, if not the war as well. "There is not *a* history," Cheyette told his colleagues at the meeting of the American Historical Association in Washington, but rather "many possible histories." Cheyette was anything but a radical. He was born in New York City in 1932 and attended Princeton after graduating from Mercersburg Academy, a private boarding school in Pennsylvania that had been founded at the end of the nineteenth century. He completed his doctorate at Harvard and, in 1963, became a professor of European history at Amherst, where he would teach for nearly fifty years. Cheyette's academic interests tended toward the conservative, as well as the more obscure corners of European history, in particular the eleventh and twelfth centuries of medieval France. In this way, Cheyette was himself a member of the academic establishment that he was seeking to challenge, and his call for reform was indicative of the broad support within the academy for dismantling the old regime of required survey courses on Western civilization—a category of history and thought whose internal coherence Cheyette and others came to believe was insufficient to justify mandatory attendance by incoming freshmen. He articulated the dominant critique of such courses at the time when he described to his academic peers "the realization that what had passed for universal was itself sectarian."

The retreat had been gathering pace for years. The first earnest challenges to the dominance of courses on Western civilization in the United States had arisen a decade before the meeting in Washington, after the convulsions of the 1960s prompted many to ask whose history was being told and taught. In some cases, as one observer recounted, the courses "died a natural death and in others were simply murdered." At Stanford, for example, the History of Western Civilization had been a required course for years after the end of World War II, introducing students to a discrete and curated selection of work, from Plato and Rousseau to Marx and Arendt. But in November 1968, a ten-person committee decided to abandon the requirement. The group, which consisted principally of academic administrators and professors, but also an undergraduate philosophy student, a nod perhaps to the democratic ethos of the moment, concluded in its report that such courses, which had been modeled on similar programs at Columbia and the University of Chicago, were "dead or dying." The world, including the United States, had been remade following the end of World War II. Only months before Stanford decided to retire its iconic survey course, Martin Luther King Jr. and Robert F. Kennedy had been assassinated. During the prior winter, North Vietnamese forces had launched the Tet Offensive against South Vietnam, which by many accounts would prove to be the beginning of the end of American involvement in the war. The dissonance between the upheavals of the decade and academia's desire to cling to what many believed was a vestige of a past that might never have existed had become too much.

The course at Stanford ended the following year, in 1969, going out, according to an article in the school's student newspaper at the time, "with a whimper and not a bang." The resistance on campus to dismantling the old regime of a required canon was muted, if not wholly disempowered, at the end. As one historian noted, by the late 1960s, once the challenge to educational requirements had gained momentum, students "encountered faculties already prepared to retreat." To many critics, the apparent arbitrariness of the editorial process of developing a syllabus for a course as ambitious as the History of Western Civilization—and selection of only a small handful of works for inclusion from such an enormous list of candidates—was alone reason to abandon the project. "We have Plato, but why not Aristotle?" asked Joseph Tussman, the head of the philosophy department at the University of California, Berkeley, in an essay published in 1968. "Why not more Euripides? *Paradise Lost*, but why not Dante? John Stuart Mill, but why not Marx?"

Such editorial disputes, however, masked the far more fundamental questions that the canon wars had exposed, and the significance of what was at stake. The survey course had flourished for decades on the premise that the American academy, along with its students, required grounding in a broader historical context, tethering the political and cultural developments of the United States to antecedents in Europe and antiquity. As a member of a faculty review board assembled by the American Historical Association in the 1890s had noted, "American history is in the air—a balloon sailing in mid-heaven—unless it is anchored to European history." The balloon, however, was now cut loose.

. . .

How did we get here? The current conception of "the West," as meaning a set of cultural and political values rooted in antiquity and extending through history to the modern era, began to take shape in the late nineteenth century. Its meaning would shift and evolve over the years, but rightly came to cohere around a set of shared practices or traditions that made possible, and indeed bearable, collective existence at a grand scale. As Winston Churchill observed in 1938, in a speech at the University of Bristol on the west coast of England, civilization "means a society based upon the opinion of civilians," that "violence, the rule of warriors and despotic chiefs, the conditions of camps and warfare, of riot and tyranny, give place to parliaments where laws are made, and independent courts of justice in which over long periods those laws are maintained." For Churchill, the rise of civilization makes possible "a wider and less harassed life" to the public.

Many have argued that the entire concept should be abandoned—that the imperfect and shifting descriptive power of "the West," if any, is overwhelmed by its historical tether to imperial theories of domination, of superiority and the subjugation of colonial subjects at the periphery of empire.* Appiah, for example, has argued in favor of abandoning the "idea of western civilisation," which for him has been "at best the source of a great deal of confusion" and "at worst an obstacle to facing some of the great political challenges of our time." The West, for Appiah and many others, became an object of moral scorn, impeding our understanding of history, burdening the task of interpretation with a cumbersome narrative architecture that obscured more than it enlightened. The edifice, they argued, must be torn down.

The deconstruction of and challenge to a monolithic and wholly

---

* Claude Lévi-Strauss, the French anthropologist, for example, lamented what he described as the "monstrous and incomprehensible cataclysm" that the development of Western civilization had brought upon the indigenous societies that were the objects of his interest—for him, that "innocent section of humanity."

coherent conception of Western civilization began in earnest in the 1960s but arguably culminated with the publication of Edward Said's *Orientalism* in 1978. Adam Shatz, the U.S. editor of the *London Review of Books*, argued in a 2019 essay, four decades after *Orientalism* was first published, that the book was "one of the most influential works of intellectual history of the postwar era." A group of critiques that had been gaining ground for years seamlessly cohered around Said's treatise, which became the vehicle through which academia would be remade.

It would indeed be difficult to overstate the power and sheer cultural force of Said's creation. The term "Orientalist" itself became an epithet of sorts among a certain swath of the ascendant cultural elite—a weapon that continues to have the ability to arrest a discussion in its tracks and a term that ironically itself became a means of constructing identity and exercising power on college campuses. As Shatz put it, the term "Orientalism," nearly half a century after its popularization by Said, "has become one of those words that shuts down conversation on liberal campuses, where no one wants to be accused of being 'Orientalist' any more than they want to be called racist, sexist, homophobic, or transphobic." The book's legacy, however, has been more complicated. One form of dogmatism, rooted in a colonial outlook, would soon be replaced by others, often similarly dismissive of competing notions of history and literature that transgressed against the new received wisdom. In the same way that the Orientalists of the nineteenth century and before had delineated certain cultures and peoples as having little to contribute, and as being less than equal to the privileged core of civilization, the academic establishment in the 1980s and 1990s would find in the wake of Said its own means of identifying and indeed *othering* certain arguments as being unworthy of critical engagement.

The book also reshaped the machinery and internal politics of humanities departments across the United States and around the world. The author Pankaj Mishra has written that *Orientalism* "launched a thousand academic careers." Indeed, the book gave birth to a new industry in American higher education, built around dismantling colonial understandings of the world, and at the same time, Mishra has argued, provided a means of self-promotion for a subset of "intellectual émigrés, largely male," who "were often members of ruling classes in their respective countries—even of classes that had flourished during colonial rule." As Mishra put it, "For a posher kind of Oriental subject, denouncing the Orientalist West had become one way of finding a tenured job in it."

The effect of *Orientalism* on the culture was so thorough and complete, so totalizing, that many today, particularly in Silicon Valley, are scarcely aware of its role in shaping and structuring contemporary discourse, as well as their own views about the world. In his biography of Said, *Places of Mind*, Timothy Brennan writes that beginning in the late 1990s, "postcolonial studies was no longer simply an academic field," but rather an entire worldview, with a highly particularized jargon, including "'the other,' 'hybridity,' 'difference,' 'Eurocentrism'"—terms that "could now be found in theater programs and publishers' lists, museum catalogs, and even Hollywood film." Indeed, a broad swath of intellectuals in the United States, and many of those adjacent to academia, including writers and journalists, situated their own politics—a politics that would emerge as the dominant form of elite establishment thinking in the United States through the 1990s and into this century, including in Silicon Valley—around a book that many would never encounter directly, and some of whom did and do not know exists.

The substantive triumph of *Orientalism* was its exposing to a broad audience the extent to which the telling of history, the act of summation and synthesis into narrative from disparate strands of detail and fact, was not itself a neutral, disinterested act, but rather an exercise of power in the world. As Said himself explained in an afterword to the book, written in 1994, "The construction of identity is bound up with the disposition of power and powerlessness in each society, and is therefore anything but mere academic woolgathering." In this way, the engine and mechanism of the production of history and anthropology were the objects of Said's study. And it was the inclination of that engine toward division, toward definition of the "us" and the "other," that for Said was itself a consequence and perhaps necessary component of the act of observation. As Said made clear, citing the British historian Denys Hay, "the idea of Europe" was "a collective notion identifying 'us' Europeans as against all 'those' non-Europeans." After nearly half a century, the observation seems unobjectionable and almost banal. But it was absolutely radical in the 1970s, destabilizing an entire academic mode of being across the university establishment. His central thesis provides the basis for much of what passes as foundational in the humanities today, that the identity of a speaker is as important if not more important than what he or she has said. The consequences of this reorientation of our understanding of the relationship between speaker and that which is spoken, storyteller and story, and ultimately identity and truth have been profound and lasting—but also, in its more extreme formulations, pernicious. It was the overextension of his principal claim that set in motion and empowered a deconstructionist movement that would, in the decades to come, successfully elevate the importance of the identity of the speaker over that which is said.*

The critics were many, and came from every angle. For one, Said seemed less interested in documenting the similar systems of "power-knowledge" that had been developed in the East to justify the subjugation of various underclasses within the subaltern world itself. As Mishra has observed, "The book displayed no awareness of the vast

---

* Some have read Said too expansively and been too aggressive in extending his central, and brilliant, idea. Said, for example, has frequently been misinterpreted as claiming that actual knowledge of the Orient was impossible. He was not a postmodernist in this sense. There were facts to be found; it was just that the motivations and ideologies of those charged with finding them needed to be exposed in order to have any hope of evaluating their work.

archive of Asian, African, and Latin-American thought that had preceded it, including discourses devised by non-Western élites—such as the Brahminical theory of caste in India—to make their dominance seem natural and legitimate."

Others attempted to hit more directly at what they perceived to be Said's central argument. William McNeill of the University of Chicago, for example, who was a defender of the Western civilization course requirements that were gradually and then more swiftly eliminated in the 1960s, had the temerity to resist the rise of what he would describe as the moral relativism that was ascendant in the second half of the twentieth century and that he and other critics claimed would often cloak itself in the more palatable rubric of multiculturalism. McNeill wrote in an essay published in 1997 that attempts to construct world history courses had themselves "often been contaminated" by what he regarded "as patently false assertions of the equality of all cultural traditions." He was not responding directly to Said, but Said and his arguments were so omnipresent at the time that anyone wading into such debates by that point was necessarily in conversation with him.

It is also a reminder of how swiftly the culture moves, given that a claim such as McNeill's would almost certainly require cancellation today. The species of historian who dared to make normative claims about culture, including the specific merits or lack thereof with respect to particular cultures, was essentially rendered extinct, or at least jobless, by the end of the twentieth century. Even modest attempts to point to the differences in economic output and military power between Europe and its former empires over the past five centuries or so have been pushed to the fringe of the cultural conversation. As the historian Niall Ferguson has observed, the principal Western empires that began their ascent in the sixteenth century came to control 74 percent of global economic production by the 1910s.

FIGURE 8
Western Empires: Share of Territory and Global Economic Output

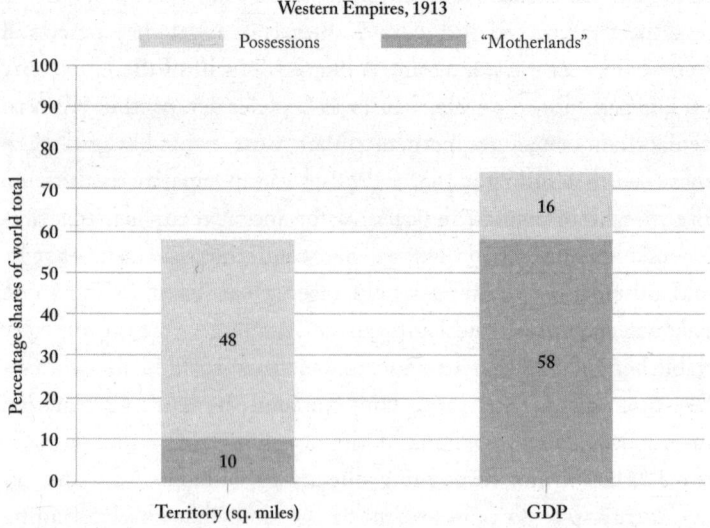

The mere recitation of such a fact has become provocative in a way that suggests our current culture's fundamental unease with truth, as well as perhaps its loss of an ability to disentangle descriptive claims from normative ones. To point out, as an empirical matter, that a certain subset of nations has come to dominate global affairs is not equivalent to the normative claim that such a result is justified. In the West, however, many observers have lost interest in investigating the causes and reasons for this outperformance. We have been taught simply to turn away, to change the subject. The ability to reckon with a descriptive claim acknowledging the overwhelming dominance of the United States and its allies while suspending discussion, even momentarily, of the moral implications of that distribution of power is arguably a form of what the journalist and opinion researcher Nate Silver has called "decoupling." This capacity for evaluating the truth of a statement while setting aside one's views about either its implications or one's opinion "on the identity of the speaker," as Silver puts it, has withered among far too many. One should be able to decide whether a descriptive claim is true without knowing anything about who is making it.

A respect for one's intellectual adversary, even if begrudging, can be an enormous advantage, particularly in a culture that has grown accustomed to belittling its opponents instead of engaging with them. In the realm of politics, and certainly business, far too many participants are incapable of maintaining a sense of emotional distance from their adversaries, of approaching them with the clarity and almost magnanimity that the best competitors bring to the arena. The most effective minds are often the ones who understand deeply the advantages and skills of their antagonists and refuse to fight religious wars of outrage and moral indignation. A fog of self-righteousness is often lethal to good judgment. As Vannevar Bush observed, writing in 1949, the failure of the Nazis to develop a sufficiently effective proximity fuse, which allowed bombs to detonate just prior to hitting their targets, was a consequence of their arrogance, not their incompetence. The Germans, he wrote, were incredulous that "the *verdammter Amerikaner*" had succeeded "where they had failed."

. . .

The systematic challenge to the West in the second half of the twentieth century, its history and identity, along with that of the American project, what it was or should be, if anything, has left a void in its wake. A regime of knowledge had perhaps rightly been torn down. But nothing has been erected in its place. The canon wars as they would come to be known on university campuses in the 1960s and later, as well as the challenge in academia to the West itself that would follow, represented a struggle not merely over the content of American identity but over whether there should be any content at all.

The thin conception of belonging to the American community consisted of a respect for the rights of others and a broad commitment to neoliberal economic policies of free trade and the power of the market. The thicker conception of belonging required a story of what the American project has been, is, and will be—what it means to participate in this wild and rich experiment in building a republic. In this country and many others, membership in the community of the nation is at risk of being reduced to something narrow and incomplete, the loose sense of affiliation that comes from sharing a language or popular culture, for example, from entertainment to sports to fashion. And many have advocated for this retreat. By the end of the 1970s, an entire generation had grown skeptical of broader national identity or shared endeavors. And that generation, including many who would go on to found Silicon Valley and spur the computing revolution, turned its attention elsewhere, to the individual consumer, disinterested in furthering the misadventures of a government whose entire project and reason for being had so thoroughly been called into question.

Chapter Eight

"Flawed Systems"

IN JANUARY 1970, *Time* magazine named "the Middle Americans" as person of the year. It was a departure from the publication's ordinary practice of highlighting a specific individual and his or her contributions on the national or international stage. After the convulsions of the 1960s, including the decade's radicalism and challenge to the reigning order, "the Middle Americans," a cohort in the metaphorical heartland of the country, far from its coasts, "feared that they were beginning to lose their grip on the country," the magazine wrote. "Others seemed to be taking over—the liberals, the radicals, the defiant young," *Time* continued. "No one celebrated them; intellectuals dismissed their lore as banality."

The same might be said today. By the early 1970s, the divide that would come to structure contemporary American politics, including the current fissures in society half a century later, had begun to open. The division of the country by *Time* into two parts, core and periphery, was an oversimplification, at best, and, at worst, a knowing appeal to a conception of American identity that predated the inclusion of a far more diverse array of minorities and immigrants. But it also captured an emerging fault line that would come to dominate American politics for decades—a divide that has only ever been loosely about policy disagreements, one that was more fundamentally concerned with culture and identity. The attacks at the time on conceptions of Western civilization, and more specifically on the internal contradictions of the American project—its claim to equality for all yet enforcement of discriminatory laws across broad swaths of the South—had only heightened the conflict. And the war in Vietnam, which seemed to have no end, along with the rise of the civil rights movement, including its direct attack on institutional complacency, had given rise to a thriving counterculture and challenge to the American establishment.

It was against this backdrop that the first glimmers of the digital revolution, of software and personal computing, and indeed artificial intelligence, took shape. The earliest collaborators and participants in the development of what would become the personal computer in the 1960s and 1970s were skeptical of government authority and had largely constructed their own identities and sense of self in opposition to the state. Lee Felsenstein, for example, who was born in Philadelphia in 1945 and later moved to Menlo Park, California, where he formed what would come to be known as the Homebrew Computer Club, one of the early groups that was focused on building prototypes of smaller computers for individual use, wrote, "We wanted there to be personal computers so that we could free ourselves from the constraints of institutions, whether government or corporate." The personal computer, as pioneers like Felsenstein saw it, was a means of liberation and emancipation from government, not cooperation with it. Stewart Brand, co-founder of the *Whole Earth Catalog*, an influential compendium for the counterculture movement of the 1960s, wrote in a 1995 essay that "the counterculture's scorn for centralized authority provided the philosophical foundations of not only the leaderless Internet but also the entire personal-computer revolution."

In the 1970s, the emerging set of technologies that would become the modern-day personal computer, as well as software more broadly, was being reinvented as a means of empowerment of the individual against the state, not a set of tools to be leveraged by the state to advance the national interest. It was an era of innovation in Silicon Valley that was driven by a mistrust of national governments, as well as frustration with their delay in adopting progressive reforms at home and their grand experiments and military misadventures on the world stage. This was not the technological revolution of Vannevar Bush or J. Robert Oppenheimer, who through much of their lives saw the purpose of technology as extending and enabling the American project. The individual, and later the consumer more specifically, would emerge as the principal object of this new industry's desire and attention.

In 1984, the author and journalist Steven Levy published *Hackers: Heroes of the Computer Revolution*, an influential chronicle of that early period of innovation in software and personal computing. Levy articulated the ethos of the moment, which was deeply skeptical of institutional and state power. "Bureaucracies, whether corporate, government, or university," he wrote, "are flawed systems, dangerous in that they cannot accommodate the exploratory impulse of true hackers," designed "to consolidate power, and perceive the constructive impulse of hackers as a threat." The human systems that government had created were too inflexible; new systems had to be built based on logic and rules instead of the capricious dictates of the elected class. The object of Levy's critique, as well as that of his confederates at the time, was a calcifying American corporate culture. Levy described the IBM of the era, for example, as "a clumsy, hulking company that did not understand the hacking impulse." The distaste for the corporate monoliths was nearly as much aesthetic as it was ethical. He continued: "All you had to do was look at someone in the IBM world and note the button-down white shirt, the neatly pinned black tie, the hair carefully held in place, and the tray of punch cards in hand." And it was the conformity of those institutions that was thought to be central to their inability to drive change.

For this emerging generation of hackers, the corporatism of postwar America and the apparatus of government were acting in concert to constrain innovation. The software and early computing devices that Felsenstein and others were building in Silicon Valley were intended to serve as a challenge to state power, not to enable it. They were not building software systems for defense and intelligence agencies, and they were certainly not building bombs.

This revolution, however, like others before it, would ultimately abandon much of its own idealism. The broader issue was that the "we" or "us" of America had so thoroughly been challenged and deconstructed—problematized, in the language of graduate school seminars today—that an entire generation of technologists turned its attention elsewhere, to the individual consumer. Steve Jobs, in particular, was a product of a waning counterculture movement in the United States, searching for purpose and direction after the conflict and storm of the 1960s began to recede. As an undergraduate at Reed College, Jobs, who would go on to lead Apple, which by some estimates would become the most valuable corporate enterprise in the history of civilization, immersed himself in a calligraphy class, where he recounted to his biographer Walter Isaacson that he "learned about serif and sans serif typefaces, about varying the amount of space between different letter combinations, about what makes great typography great." His immersion in letterforms was not a detour from his core, animating interests. It was a result of them. Jobs continued: "It was beautiful, historical, artistically subtle in a way that science can't capture, and I found it fascinating." This blend of artistry and engineering would become the hallmark of Jobs's design sensibility and was, for Isaacson, "yet another example of Jobs consciously positioning himself at the intersection of the arts and technology." To be clear, Jobs was a radical and creative savant who saw the future and made it real. His ambition was to remake the world, not tinker at its margins. When attempting to court John

Sculley, then president of PepsiCo, to persuade him to join Apple as chief executive officer, Jobs reportedly asked him, "Do you want to spend the rest of your life selling sugared water, or do you want a chance to change the world?"

Jobs's revolution, however, was essentially intimate and personal. His principal focus was on constructing the products—including the mobile phones that now coexist with us on our person throughout our lives—that would liberate the individual from reliance on a corporate or governmental superstructure. And he did. His interest was not in building the means to advance a broader American or national project, or in enabling a closer collaboration between the technology industry and the state. Indeed, Apple objected to attempts by the U.S. government, including the Federal Bureau of Investigation, to unlock its iPhones in connection with investigations in criminal cases. The products that Jobs and Apple built were focused on the power and creativity of the individual mind and as a result were extensions, often literally—in the form of phones, wristwatches, personal computers, and the mouse—of the self.

For Apple in the early 1980s, the personal computer presented a challenge to, not an embrace of, the authority of government and the state. The company's iconic "1984" advertising campaign featured a dystopia of conformity, filled with hundreds of gray souls mindlessly listening to the directives of an Orwellian overlord speaking to the assembled flock on a large screen. A woman, dressed in bright tangerine running shorts, sprints through the crowd and throws a sledgehammer at the screen, smashing it and, for the viewer, suggesting the liberation of the masses. The television ad, directed by Ridley Scott, pitted the emancipatory potential of the Macintosh computer against the then reigning IBM, which had produced the gigantic mainframe computers of an earlier generation that often literally filled entire rooms.

Those mainframes, hulking and immovable, would only, Apple implicitly argued, hasten our enslavement by the state. The Macintosh, by contrast, weighed seventeen pounds and had a handle on the top, so that it could literally be picked up and carried short distances by a single person. An initial draft of the advertisement warned ominously that "there are monster computers lurking in big business and big government that know everything from what motels you've stayed at to how much money you have in the bank." The message was clear: the new personal computer of the era would provide a counterweight to the institutional power of government and industry, not advance their interests at the expense of the individual.

Our point is only that the rush of attention and funding dedicated to the concerns and needs of the modern American, and later global, consumer was anything but inevitable. It was the product of a set of proclivities and instincts of those early founders, as well as the social and cultural milieu in which they came of age. They had ambition, no doubt. But much of their focus was on the individual, his and her concerns and needs. And it was a near-obsessive focus on those concerns and needs—and the sheer brilliance of the contraptions and software products constructed to address them—that paved the way for another generation of founders, in the first part of this century, who would create the consumer internet. The era of online advertising, photo-sharing apps, and food delivery empires was near. This next generation of innovators would go even further than the prior, abandoning even the pretext of claims to a broader political project, to the liberating potential of technology. They instead entered into a far more mercenary and straightforward service of the material culture of the time.

Chapter Nine

Lost in Toyland

IN 1996, TOBY LENK, a vice president of corporate strategic planning at the Walt Disney Company, was offered a job leading the entertainment giant's theme park division—the iconic group that had opened Disneyland, in California, in 1955, and later Disney World, in Florida, in 1971. He, along with hundreds of others across the American corporate landscape at the time, however, was captivated by a different magical realm: the internet, and the inroads the technology had made into homes and consumer culture. Lenk, who was born in Framingham, Massachusetts, and had earned an MBA at Harvard Business School after attending college at Bowdoin, decided to leave the relative safety of the Disney empire to found his own company, selling toys on the internet.

The company, eToys, was for a brief moment the envy of much of Silicon Valley. At its height, the company's market capitalization reached $10 billion after its IPO in 1999, only two years after its founding. Lenk himself might have been worth $850 million at one point. For many investors searching for their next wager, he "stood out as a grown-up" in the startup space, "at a time when Wall Street money was cascading down on barely postpubescent entrepreneurs," as one journalist put it. The surge of interest from an emerging venture capital community, and later the broader public, was unrelenting. It was clear to everyone that a historic shift in the way commerce would be conducted had arrived. And the race to begin selling goods online had begun.

Lenk's pitch was anything but contrarian in light of the prevailing mood among startups at the time. "We're losing money fast on purpose, to build our brand," he told *Advertising Age* in an interview in June 1998. For some, the unrepentant abandonment of the old rules of business, including the inconvenient requirements of traditional accounting and the goal of profitability, exposed the hubris of this new rush of founders. Others, however, appreciated the new time horizon that they, and their ventures, sought to embrace. The arrival of the internet had upended global commerce, and the effects of that shift would be revealed not over months or years but over generations. The time for investment, and perhaps losses, was now. The approach of eToys was nearly identical to that of the flood of other, similar startups—from Pets.com (pet supplies) to Boo.com (clothing) to Kozmo (groceries and video games)—racing to monetize the shift of shopping to the internet. Take the market, first. Profits, second. An estimated fifty thousand companies with $256 billion in funding were formed at the height of the growing bubble.

The appeal of eToys was that its business model did not require much imagination. As a *Wall Street Journal* profile of the company noted during its ascent, "To a person searching for wooden trains, eToys seems like an online version of the corner toy shop; to a person hunting for the latest 'Star Wars' paraphernalia, it's a giant toy store without the battling crowds." It was clear to everyone, including the investing public, the benefits of moving sales of toys online. In May 1999, in its S-1 filed with the U.S. Securities and Exchange Commission prior to its IPO the following month, eToys outlined the current friction in the shopping experience for many parents, listing twelve steps involved, including "circle parking lot 4 times for parking space," "lose one child in the Barbie section," "drive home," and, belatedly, "remember you need gift wrap." There were some skeptics, but Lenk was unconcerned. "There is all this talk about

Toys 'R' Us and Wal-Mart, blah, blah, blah," he said in 1999, with characteristic confidence if not bluster. "We have first mover advantage, we have defined a new area on the Web for children. We are creating a new way of doing things." His rhetoric mirrored that of a new breed of founder and heralded a new type of investing, focused not on marginal growth but on the aggressive disruption of incumbents and the construction of new monopolies.

For all of its ambition and revolutionary rhetoric, eToys was, still, a toy company. It was squarely focused on the consumer, and the business proposition was anything but ornate—sell more of the same thing through a different channel. Our critique here is not that the pursuit of consumer markets is misplaced but rather that such a single-minded focus on the consumer has come at the expense of other broader and more significant endeavors. We do not intend to fetishize a nonmaterial existence, casting consumption and objects of desire as the enemy to purity and enlightenment. To desire, even a toy, is to be human. To want is to situate oneself within the world. In a particularly intimate scene from *Before Sunset*, the second film in Richard Linklater's iconic three-part meditation on romance, with Julie Delpy and Ethan Hawke, the two actors, the archetypical flaneurs perhaps, stroll through the streets of Paris on a sunny afternoon, over the course of a playful and meandering conversation. Hawke's character, Jesse, offers the familiar challenge to the traps of consumption and to materialist desire. "I just feel like I'm designed to be slightly dissatisfied with everything," he says, wistfully. "I satisfy one desire, and it just agitates another." Celine, played by Delpy, responds, winning the exchange: "But I feel really alive when I want something. . . . Wanting, whether it's intimacy with another person or a new pair of shoes, is kind of beautiful. I like that we have those ever-renewing desires."

The issue with eToys and others was not their interest in sating our wants or needs. It was the shallowness of their ambition and abdication of everything beyond the light hedonism of the moment.

The energy of the era was directed at addressing the inefficiencies that would-be founders encountered in their own quotidian lives; it empowered a certain type of excavation of the problems of modern life, which against the backdrop of a broad and essentially successful challenge to any sense of a national project had become oriented around material culture. Everyone could be a founder, because everyone encountered things that needed fixing and better ways to navigate their daily lives. This democratization of the potential for producing novel ideas in business, to challenge incumbents, has been one of the most enduring effects of the rise of the consumer internet, its websites, and the avalanche of apps. Lenk, for example, told an interviewer in 1999 that he had additional business ventures that he was considering pursuing. "I'm a keen golfer and there were no places that you could practise if you weren't a member of a private club, no place to putt," he explained. "I was going to try to create high-quality practice greens for the public." His pitch of putting greens for the people was emblematic of the era. The excess of capital and lack of any broader or unifying collective project to focus the entrepreneurial energy that had been unleashed across the country had left founders turning inward, to address their own personal challenges, however idiosyncratic, which often meant managing the inconveniences and indeed indignities of daily life.

There was almost too much to *disrupt*. The term itself would eventually be robbed of real meaning. The era of the casual founder, of indiscriminate disruption, had arrived. An initial cycle of genuine creation, built on the back of a novel technology that was capable of connecting every computer on the planet, had begun to degrade into something derivative. The artist Jean-Michel Basquiat, whose paintings in the 1980s demanded that the boundaries of what could be considered high art be redrawn, incorporated elements of graffiti and street art in his work. So much of what made his paintings original, however, would later be repurposed and recycled, almost endlessly, by a culture ravenous for even hints of the novel. Some of that borrowing and reassembly is itself new and fresh. But much is not. The same was true of the heady days of the rise of the young internet in the late 1990s. There was some real art being created, some Basquiats refining their craft. But most of the companies were lifeless and derivative works.

* * *

For a later generation of founders, beginning in the 2010s and continuing today, the inconveniences of daily life for those with disposable income—hailing taxis, ordering food, sharing photos with friends—would eventually provide much if not most of the fodder for their inventions. The entrepreneurial energy of a generation was essentially redirected toward creating the lifestyle technology that would enable the highly educated classes at the helms of these firms and writing the code for their apps to *feel* as if they had more income than they did. The cognitive dissonance for this generation was severe. They had the cultural and educational pedigree of an aristocracy but not the bank account. These were not the hereditary elites and blue bloods of a prior era. This was a new coalition, the product of America's vaunted meritocracy and radical experiment to essentially throw open the doors to its most hallowed educational institutions to a new swath of talented young minds. But as Peter Turchin has argued, in his book *End Times*, the unintended result of the country's focus on higher education, as opposed to birth or caste, as the new means of constructing an overclass was an "overproduction" of elites that created too many qualified candidates for too few jobs.

The frustrations and resentments of those who perceive themselves to have been deprived of opportunities to which they are entitled can overwhelm the most resilient minds. Talcott Parsons, the American sociologist who was born in Colorado Springs in 1902, has argued that the majority of adult men are "condemned to what, especially if they are oversensitive, they must feel to be an unsatisfactory experience," deprived of their rightful inheritance. Parsons was the last of a generation of theoretical sociologists whose work was unencumbered, or as critics would charge, uninformed, by empirical research.* His insights, however, were often all the more penetrating. In an essay on human aggression published in 1947, Parsons observed that many men "will inevitably feel they have been unjustly treated, because there is in fact much injustice, much of which is very deeply rooted in the nature of the society, and because many are disposed to be paranoid and see more injustice than actually exists." And he went further. The feeling of being "unjustly treated," Parsons noted, is "not only a balm to one's sense of resentment, it is an alibi for failure."

The creative energies of Silicon Valley engineers would end up being directed toward solving their own problems, which, for many, stemmed from a fundamental disconnect between the life they thought they had been promised as a result of their intellectual talents—a life of ease and wants sated, of car services and assistants at the ready to fetch meals and groceries—and the reality of their relatively modest incomes. This generation was told that they were bound to become the next masters of the universe, but there was little for them to inherit. So they would ultimately go about constructing the apps and consumer services that would create an illusion of the good life for themselves and their peers by making it possible to summon taxis, make restaurant reservations, and book vacation home rentals with only a few swipes on a phone.

The initial bubble of the late 1990s, of course, would ultimately burst. After sales at eToys lagged, the market grew increasingly impatient. In February 2001, the company's shares traded for a mere nine cents a piece, after having reached a high of $85 only a few

---

* An essay in *Commentary* magazine from 1962 noted that Parsons, in "an intellectual milieu dominated by empiricists," had "been able to 'get away with' (as he put it once, in an unusual moment of irony) Pure Theory."

years before. eToys filed for bankruptcy that month. An entire generation of consumer internet startups was washed away in the reckoning. "A year ago Americans could hardly run an errand without picking up a stock tip," an editorial in the *New York Times* stated on Christmas Eve in 2000. "What a difference a year makes." The newspaper noted that eToys, for example, along with Priceline and many other "former Wall Street darlings, have seen their stock prices fall more than 99 percent from their highs." For his part, Lenk blamed the excesses of the moment, "this craziness, this frothing," as he later described it, for his company's fall from a quite fleeting grace. The conventional wisdom was that the capital markets, along with venture capitalists, were the principal culprits behind the collapse. In a postmortem of the crash published in May 2001, D. Quinn Mills, a professor at Harvard Business School, wrote that "traditional business plans and financial measures didn't apply" to this new breed of startup. "Yet investors continued to use the old tools, pressuring start-ups for impossible specificity in their strategies and reckless speed in implementing them," he added. The confluence of factors in driving the euphoria of the moment had been historic. The *Guardian* noted at the time, from its arguably more neutral vantage across an ocean in the United Kingdom, that "the mania for technology stocks" had "all the ingredients for a roller-coaster ride from boom to bust—glamorous sounding products that investors knew little about, avarice, an economy firing on all cylinders, some dashing young entrepreneurs, a small army of cheerleaders in brokerage houses and in the media peddling the line that the rules of business had been rewritten." The chapter had ended, and many in Silicon Valley were simply in awe of the scope and extent of the destruction.

The criticism of that early generation of startups focused on their lack of discipline and reckless spending, as well as the abandonment of any rigor or scrutiny from their investors. But there was a far more fundamental misallocation of resources, of capital and talent. The failing of that early internet era was its rush to serve the needs of the consumer at the expense of those of the nation-state or public. And that focus on the consumer endures to this day. The lack of ambition from many startups today is and remains striking. Far too much capital, intellectual and otherwise, has been dedicated to sating the often capricious and passing needs of late capitalism's hordes. Others have raised similar critiques. As David Graeber wrote, "Where, in short, are the flying cars? Where are the force fields, tractor beams, teleportation pods, antigravity sleds, tricorders, immortality drugs, colonies on Mars, and all the other technological wonders any child growing up in the mid-to-late twentieth century assumed would exist by now?" His interest was in disentangling the structural causes of the West's failure to fulfill the promise of its own mythology of unrelenting scientific and technological progress. For Graeber, who described himself as an anarchist, the technology industry, and American culture more broadly, were at risk of being reduced to a sort of technical "pastiche"—the rearrangement and repurposing of existing content and breakthroughs. The end of innovation was perhaps coming into sight. The apps and games and video-sharing platforms that were being built en masse, that were consuming enormous sums of money and talent at the expense of more significant projects, were anything but idle and innocuous diversions. And the lasting effects and harm of this new form of screen-based competition for our attention, particularly on children, have only begun to be unraveled.

* * *

At a gathering of lobbyists and economists in December 1996 in Washington, D.C., Alan Greenspan gave a speech in which he issued his now famous warning of "irrational exuberance" in the markets. The remark has come to define that particular moment of excess and spawned an entire industry of research and ongoing debate. But the investors hoarding shares of this new generation of companies were not wrong. They were just early. A small number of the startups from the era, including Amazon, Google, and Facebook, would go on to become some of the most dominant commercial enterprises in the world. The exuberance of the time had been not so much irrational as indiscriminate. Entire sectors, including enterprise software and defense and intelligence systems for the military, had also been overlooked in the rush to reimagine online shopping. There were vast expanses of opportunity that had been passed over by the wisdom of the crowds and the market.

Silicon Valley had made clear its disinterest in the work and challenges of government. The barriers to entry were too high, the budget cycles too long, and the politics too messy. But a wave of founders had, perhaps unintentionally, stumbled on something even more valuable than the software they were building: a new organizational culture and means of marshaling the talents of individuals. Many of the businesses were rightly swept aside. But it was the organizational culture that was left amid the economic wreckage, an engineering mindset that constituted a new approach to channeling the efforts of a group, that might have been the era's most enduring and transformative product.

Part III

The Engineering Mindset

Chapter Ten

The Eck Swarm

ON JUNE 26, 1951, at around 1:30 p.m., a cluster of honeybees began to form in a park in Munich, Germany. This small swarm of bees would eventually help reshape our understanding of the animal mind and its capacity for undirected cooperation. Martin Lindauer, a researcher at the University of Munich's Zoological Institute, was on hand that summer afternoon to document the swarm as part of a study on the behavior of the hive and the ability of bees to coordinate among hundreds and even thousands and tens of thousands of individuals. He was captivated by the behavior of the species *Apis mellifera* and was determined to shed light on the delegation of responsibility among individuals within a single bee colony, particularly when they began searching for, and deciding between, new potential nesting sites.

Lindauer was born in 1918 in southern Bavaria. His father, who kept beehives as well, was a farmer, and the family had fifteen children. As Hitler rose to power and war engulfed the continent, Lindauer ended up serving in the German army for three years. His interests, however, lay elsewhere, and after suffering an injury on the Russian front in 1942, he was discharged from the military. Thomas D. Seeley, a biology professor at Cornell University who has written extensively on Lindauer's work, has noted that Lindauer once described the scientific community to which he would return after his time in the army as "a new world of humanity." The exploration of the natural world was a reprieve for Lindauer, who retreated into science after the war.

He was part of a generation of zoologists whose work preceded the rise and eventual dominance of genetic-based research in the field. For a time during the nineteenth and twentieth centuries, the best access that biologists such as Lindauer had into the mind of the animal was through its outward behavior; a more complete understanding of the power and inner workings of the gene, as a means of accessing the nature of a species, was still out of reach. These earlier generations of scientists of the natural world, including the French psychologist Alfred Binet, were observers in the field, and keen ones. The mysteries underlying the behavior of the animals and humans that they were studying, invisible to most, were there for the taking, at least to anyone who was capable of looking closely and for a sufficiently long time.

When animals search for a new home, whether geese, leaf-cutter ants, horses, or sparrows, they often venture out as single individuals, and sometimes in pairs, in search of suitable accommodation. The practice of the European honeybee, however, departs significantly from the norm. Whereas most animals explore their environments independently, in the case of honeybees "a large community of 20 to 30 thousand individuals *together* move into a new nest-site," Lindauer wrote—a process that requires immense coordination but without a central queen bee or other specialized leaders directing the work of the group. The process by which tens of thousands of individual organisms manage to organize themselves, canvass potential nesting sites, ultimately select one of a number of options over the rest, and then together move to their new home was an absolute puzzle to Lindauer and his contemporaries.

On this particular summer afternoon in 1951, the collection of bees that Lindauer had been watching was small at first. They had begun congregating not far from an imposing stone statue of Neptune, holding a trident and rising from the waters of a nearby fountain. The University of Munich's Zoological Institute, which had granted permission to Lindauer to study the bee colonies that the institute maintained, was located in a park that had served as the site of a botanical garden constructed in the early nineteenth century, and there were plenty of secluded and attractive potential nesting sites nearby among the trees and foliage. At around three that afternoon, clouds began forming over the park, at which point Lindauer noted that the bees retreated to a nearby bush, where they stayed and spent the night. The following day, after the cloud cover broke and the sun returned, the bees resumed their work of searching for a place to build a hive.

Such searches were involved affairs. They included dozens and sometimes hundreds of scouting bees canvassing potential options nearby. The bees return to the group and perform what has come to be known as what Karl von Frisch, an Austrian-born zoologist and colleague of Lindauer's who would later win a Nobel Prize for his work on the subject, described as a dance language, or *Tanzsprache*— a method of communication by the bees that involved rocking their bodies back and forth in front of onlookers that would gather to watch. Frisch and Lindauer had discovered that the distance of this dance, that is, whether the scouting bee walked for a centimeter or two, for example, was proportional to the distance of the potential nesting site from which they had returned, and therefore indicated how far of a flight it would be to get there. In addition, evidence had begun to accumulate suggesting that the angle of the walk, relative to the position of the sun, indicated the direction of the new nest site. Over the course of the afternoon, scouting bees had returned to the main swarm to report eight potential nesting sites in the area, including a crack at the molding on top of a nearby window, a woodpecker hole, and a small hollow in a tree. It had become evident to Frisch and Lindauer that individual scouting bees would perform dances in favor of different sites and that the number of scouts that danced in favor of various locations would allow the hive to essentially vote as to the best option.

The bees, for Lindauer, represented something different in nature.

FIGURE 9
Locations of Potential Nesting Sites as Indicated by Honeybee Dances in the Eck Swarm

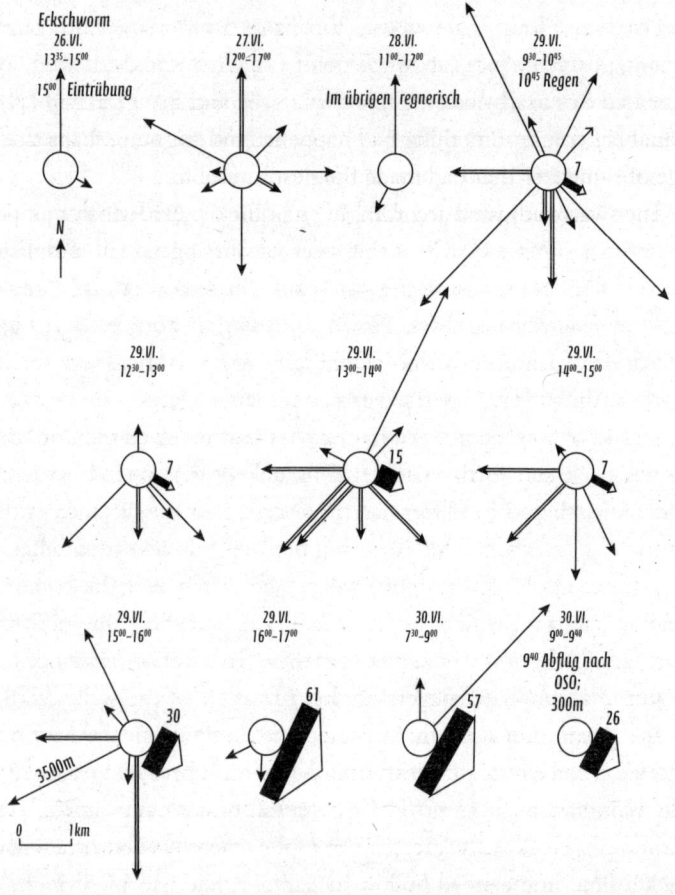

The swarm that he was observing was not merely a collection of discrete individual animals. The precision and extent of their coordination, and lack of any apparent means of centralized management, made clear that the bees formed a discrete system, a coherent whole, whose capacity for assessing and adapting to its surroundings would prompt a reassessment of what constitutes an organism in the decades to come. Lindauer narrated the scene with a blend of delicacy and reverence, noting that while two of the eight sites "had already received somewhat more popularity," "naturally there was not yet any talk of an agreement." On the following day, he noted that the scout bees had seemingly become less enthused about the north site, presumably because something had happened overnight, perhaps a deluge of rainwater that had made the nest unusable.

The swarm adjusted accordingly, and quickly. A new batch of potential sites was located by the scouts, some of which, Lindauer wrote, "were only announced by a single dance and received no attention from the population at large," while "others were lavished with more attention." Over the next several hours, the bees continued to dance in favor of their preferred nesting sites—a blur of intensity and movement through which a collection of thousands of individuals were negotiating and ultimately voting on their top contender for a new home. A particular spot three hundred meters away eventually "emerged as the favorite," Lindauer reported. The remaining holdouts had relented and given in. The following morning, at 9:40 a.m., Lindauer observed that the entire swarm of bees, having negotiated over the options and settled on a preferred location, "took off and moved into its new home."

The observations of the Eck Swarm, as it would come to be known, represented a critical moment in our understanding of the behavior of honeybees and their capacity for communication.* But Lindauer's work also suggested something more fundamental about the ways in which groups, and indeed extraordinarily large groups of individual animals, have the potential to organize themselves around

---

* The names of the swarms under observation in many cases came from their locations in Munich (e.g., the "fence," "elm," and "hedge" swarms). The German word *Eck* means "corner" in English.

a particular problem and respond to changing conditions. As one group of researchers has noted, writing on the implications of the collective decision making of honeybees and other animals for human organizations, including nurses and physicians in the healthcare field, the social structure of bees demonstrates "coordinated behaviour that emerges without central control."

The startup, in its ideal form, should become a honeybee swarm. Such coordination and movement, without an overbearing and unnecessarily centralized mechanism of control, is in many ways the single most essential feature of successful startup and engineering cultures in the American context. The bees that Lindauer and others since have studied do not incorporate caste-based social hierarchies in order to address the enormous collective action challenges that they face, but rather distribute autonomy to as great a degree as possible to the fringes—the scouts—of their organization. The individuals at the periphery of a group, who often have the latest and most valuable information regarding the suitability of potential nesting sites, and can take into account shifting conditions, are the ones who cast their ballots by dancing for the group. The swarm *organizes itself* around the problem at hand.

Other species have demonstrated similar patterns of behavior. Giorgio Parisi, an Italian physicist, has studied starlings for years in the hope of understanding the means by which they pass information to one another so quickly and are thus capable of flying in the whirls of flocks that seem to move as a single unit. In December 2005, he and his team set up three cameras on the top of the Palazzo Massimo, a building in central Rome that houses the National Roman Museum. Each of the cameras was set to photograph the flocks of starlings that routinely hovered and whirled above the square, taking a total of ten images every second. He found that the flocks of birds, which to casual observers are often thought to be spheres or oddly shaped orbs, are actually more like disks. With his ten images every second, and a three-dimensional reconstruction of the birds moving through space, Parisi's team was able to map the precise position of each bird in a given flock.

As is the case with the honeybees, the movements of the group of starlings are most often initiated by birds at the edges of the flock, those with a best vantage of potential predators and the world outside—not by preordained leaders or chiefs. Guidance as to which direction the group will be moving is then passed from bird to bird, from the edges of the flock to its core, within a fraction of a second, and shared seamlessly across the entire group of hundreds of individuals. As Parisi wrote, messages regarding which way to fly among birds in the flock are shared among them "as if by incredibly rapid word of mouth."

. . .

At most human organizations, from government bureaucracies to large corporations, an enormous amount of the energy and talent of individuals is directed at jockeying for position, claiming credit for success, and often desperately avoiding blame for failure. The vital and scarce creative output of those involved in an endeavor is far too often misdirected to crafting self-serving hierarchies and patrolling who reports to whom. Among the bees, however, there is no mediation of the information captured by the scouts once they return to the hive. And the starlings do not have to seek permission from higher-ups before they signal to their neighbors that the flock is turning. There are no weekly reports to middle management, no presentations to more senior leaders. No meetings or conference calls to prepare for other meetings. The bee swarms and flocks of starlings do not consist of layers upon layers of vice presidents and deputy vice presidents, directing the work of subgroups of individuals and managing the perceptions of their superiors. There is only the flock or the swarm. And it is within those whirls of motion that a certain type of improvisation, and looseness, is allowed to take form.

Chapter Eleven

The Improvisational Startup

FOR YEARS, NEW EMPLOYEES AT Palantir were given a copy of a somewhat obscure book on improvisational theater published in the late 1970s by Keith Johnstone, a British director and playwright. Johnstone is credited with articulating much of the theory underlying improv, as it has come to be known in the United States—an approach to acting that has in many ways overtaken the contemporary understanding of humor in film and television culture. The volume is slim and seemingly unrelated to computer science or building enterprise software. New employees were often surprised to receive it.

The parallels, however, between improvisational theater and the plunge into the abyss that is founding or working at a startup are numerous. To expose oneself on the stage, and to inhabit a character, require an embrace of serendipity and a level of psychological flexibility that are essential in building and navigating the growth of a company that seeks to serve a new market, and indeed participate in the creation of that market, rather than merely accommodate the needs and demands of existing ones. There is a breathless, improvisational quality to building technology. Jerry Seinfeld has said, "In comedy, you do anything that you think might work. Anything." The same is true in tech. The construction of software and technology is an observational art and science, not a theoretical one. One needs to constantly abandon perceived notions of what *ought* to work in favor of what *does* work. It is that sensitivity to the audience, the public, and the customer that allows us to build.

Johnstone's book also reveals one of the principal features of modern corporate culture that arguably inhibits the growth of an engineering mindset—the essential feature of an insurgent startup. He was born in 1933 in Devon, England, along the country's southwest coast. His father was a pharmacist, and the family lived above its drugstore downstairs. In *Impro: Improvisation and the Theatre*, which was first published in 1979 and has evolved into something of a cult classic among students of improvisational comedy, Johnstone blends a discussion of acting and human psychology as he reviews various exercises that he has used in his theater workshops with aspiring actors and improvisational comedians. His discussion of status, by which he meant the relative power relationship between two individuals in a given context, is particularly relevant for building flexible engineering cultures that are focused on outcomes as opposed to merely constructing and inhabiting elaborate and self-serving hierarchies. One of his central insights is that status, like other character traits, is in many ways *played*, and that actors and improvisational comedians can elevate their craft by acquiring and refining a sensitivity to what Johnstone refers to as the status transactions and negotiations that result when two individuals encounter each other in the world. In the context of a lesson on acting, for example, he observes that subtle gestures and signals between two people onstage—such as an aversion to eye contact, a nod of the head, or an attempted interruption by one actor of the other—are all methods of negotiating and asserting status relationships relative to one another. The point is that stature, in the world or on the stage, is anything but fixed or innate. Rather, it is best thought of as an instrumental attribute or good—one that can, indeed must, be wielded in service of something else.

Johnstone's interest and approach to status, and to exposing the often invisible pecking orders around us, were influenced by the work of the Austrian zoologist Konrad Lorenz, particularly his 1949 book, *King Solomon's Ring*, a collection of observations on the social behavior of various animals, from jackdaws, a relation of ravens and crows, to wolves. The most dominant jackdaws, for example, are particularly dismissive of the bottom rungs of their flocks, Lorenz tells us, so much so that "very high caste jackdaws are most condescending to those of lowest degree and consider them merely as the dust beneath their feet." The same could be said of the rigidity of internal cultures within a traditional business, with layers upon layers of hierarchy preventing ambition and ideas from rising to the top. For Johnstone, "every inflection and movement implies a status," and "no action is due to chance, or really 'motiveless.'" In particular, a bifurcation of the "status you are and the status you play," as Johnstone put it, is essential to maneuver effectively on the stage and in the world—to not be limited by the attempts of others to constrain one's freedom of movement from a business or social perspective, or at a minimum to become more aware of those attempts at domination and to respond accordingly. One can also more readily identify pockets of talent and motivation within an organization once the veil of status, the constricting gauze through which everything is perceived in corporate life, is lifted.

The broader difficulty of traditional American corporate cultures is that they tend to require a union of the status that one *is* and the status that one *plays*, at least with respect to the internal forms of social organization within the business. The senior executive vice president at a company, for example, is too often a senior executive vice president in all contexts and for all purposes internally, and his or her rank with respect to others requires an unwavering dominance in areas where such dominance may or may not advance the goals of the institution. A turn toward more rigidity and structure within American businesses gathered pace after the end of World War II. By the 1960s, for example, the electronics manufacturer Philco, which was founded in 1892, had created an ornate internal hierarchy with accompanying rule books that specified the type of furniture executives were allowed to have in their offices based on their seniority within the company. This level of rigidity in internal social structure falls far, of course, from Lindauer's swarm.

Along the lines of Johnstone's *Impro*, we have, at Palantir, attempted to foster a culture in which status is seen as an instrumental, not intrinsic, good—something that can be used and deployed in the world to accomplish other goals or aims. A significant misconception of not only the organizational culture of Palantir but many other companies with roots in Silicon Valley is that such companies have flat or no hierarchies. Every human institution, including the technology giants of Silicon Valley, has a means of organizing personnel, and such organization will often require the elevation of certain individuals over others. The difference is the rigidity of those structures, that is, the speed with which they can be dismantled or rearranged, and the proportion of the creative energy of a workforce that goes into maintaining such structures and to self-promotion within them.

We undoubtedly have some form of "shadow hierarchy" within the company, power structures that are not telegraphed explicitly but exist nonetheless. The lack of organizational legibility comes at a cost, increasing the price of navigation internally, for employees, as well as for outside partners, who often simply want to know who is in charge. But many discount the amount of open space that a deemphasis on internal signs and signifiers of status, for thousands of employees, can create. The benefit of it being somewhat unclear or ambiguous who is leading commercial sales in Scandinavia, for example, is that maybe that someone should be you. Or what about outreach to state and local governments in the American Midwest?

The point is only that voids or perceived voids within an organization in our experience have repeatedly had more benefits than costs, often being filled by ambitious and talented leaders who see gaps and want to play a role but might otherwise have been cowed into submission for fear of venturing onto somebody else's turf.

. . .

At many large companies across the United States and Europe, and around the world, it is now commonplace to routinely hold meetings of twenty, thirty, even fifty or more people on a weekly basis, and sometimes multiple times per day. More often than not, however, these gatherings are merely mechanisms through which corporate elites jockey internally for stature and resources. The faux presentations and talking points merely serve to advance the interest of politically talented, but often substantively less valuable, personnel whose principal contribution to the output of the corporation can be vanishingly hard to measure. These lengthy meetings are often preceded by even more internal pre-meetings, where employees prepare to meet with one another.

The meeting-industrial complex has driven some toward the edge and, apparently, even self-harm. A group of researchers at Harvard Business School interviewed 182 executives across industries, from the tech sector to consulting, and found a widespread feeling of being overwhelmed, suffocated by the volume and duration of meetings in contemporary corporate culture. One executive even confided that she had resorted to "stabbing her leg with a pencil to stop from screaming during a particularly torturous staff meeting." Such meetings are mechanisms by which the ambitious self-promoters within an organization telegraph their status and power, and many talented but less manipulative colleagues simply choose to relent, at a significant cost to the institution.

The principal limitation of contemporary corporate cultures is that the hierarchies and social organization of companies are far too rigid to accommodate new and shifting challenges. In January 1988, Peter F. Drucker, the management theorist whose work gave rise to an entire field of scholarship on the inner workings of large institutions, from General Electric to IBM, published an essay in *Harvard Business Review* that argued a new model of management would soon come to dominate American businesses and large organizations. It was prescient. A symphony orchestra, for example, should, based on the prevailing conceptions of how organizations ought to be structured, have "several group vice president conductors and perhaps a half-dozen division VP conductors." Orchestras, however, had no such layers. As Drucker explained, "There is only the conductor-CEO—and every one of the musicians plays directly to that person without an intermediary. And each is a high-grade specialist, indeed an artist." Drucker's central insight was that a direct line of contact—and indeed eye contact, in the case of an orchestra conductor—between a corporate leader and the creative producers within his or her organization is essential. And in our experience, the most talented software engineers in the world are artists, no different from painters or musicians. An unnecessarily structured organization alienates such talent from the goals of the institution at an enormous cost.

The flaw, and indeed tragedy, of American corporate life is that the vast majority of an individual employee's energy during their working lives is spent merely on survival, navigating among the internal politicians at their organizations, steering clear of threats, and forming alliances with friends, perceived and otherwise. We and other technology startups are the beneficiaries of the sheer exhaustion that many young and talented people either experience or can sense from the American corporate model, which can be an unapologetically extractive enterprise that too often requires a redirection of scarce intellectual and creative energy toward internal struggles for power and access to information.

In this way, the legions who have flocked to Silicon Valley are cultural exiles, many of whom are extraordinarily privileged and empowered, but misfits and thus exiles nonetheless. They have consciously chosen to remove themselves from capitalism's dominant corporate form and join an alternative model, imperfect and complex, to be sure, but one that at its best suggests a new means of human organization. The challenge, in this country and others, will be to ensure that the most talented minds of our generation do not splinter off and form their own subcultures and communities separate and apart from the nation. The homes that they find must be incorporated into the whole.

. . .

We have over the past century essentially cast culture aside, dismissing it as overly specific and exclusionary. But in Silicon Valley—even as many have neglected national interests—a set of cultural practices has proven so generative of value, that we ought to take them seriously, and particularly as ideas that might provide a basis for rethinking our approach to government, and the provision of public services. Why should the private sector alone be the one to benefit? Many seem to be watching the rise of Silicon Valley at almost a distance, eager, of course, to make use of the contraptions and services that it has produced and occasionally indignant at the industry's concentration of power, but essentially observing from afar. Where is the desire and urgency to co-opt and incorporate the cultural values that are the precondition for what the Valley has been able to build? One of the most significant mistakes made by observers of the technology industry's rise is to assume that the software produced by such companies is the reason for their domination of the modern economy. It is rather a set of cultural biases and practices and norms that make possible the production of such software, and thus are the underlying causes of the industry's success.

The central insight of Silicon Valley was not merely to hire the best and brightest but to treat them as such, to allow them the flexibility and freedom and space to create. The most effective software companies are artist colonies, filled with temperamental and talented souls. And it is their unwillingness to conform, to submit to power, that is often their most valuable instinct.

Chapter Twelve

The Disapproval of the Crowd

In 1951, SOLOMON E. ASCH, a professor of psychology at Swarthmore College in Pennsylvania, conducted a seemingly straightforward study on the human inclination to conform when faced with pressure from a group—an experiment that would prompt a far broader reckoning with the fragility of the human mind. And it was one of a number of studies in the early postwar period that captured an essential feature of our psychology that must be overcome in order to construct a company from scratch.

Asch was born in Warsaw in 1907, in what was then the Russian Empire. When he was thirteen years old, his family immigrated to New York, where he attended City College and later earned his doctorate at Columbia University. In his conformity experiments, which exposed to a broad audience the limitations of human willpower to resist the pressure of the group, Asch arranged for an instructor in a classroom to show placards with a control line, alongside three additional lines of varying heights, each of which was numbered, to a group of eight individuals, only one of whom was a true test subject. The other seven were confederates of the experimenter. Each of the eight participants was then asked which of the three numbered lines was the same length as the control line. In the following example, the correct answer would be line 2, which matches the length of the unnumbered control line on the left.

FIGURE 10
The Asch Conformity Experiment

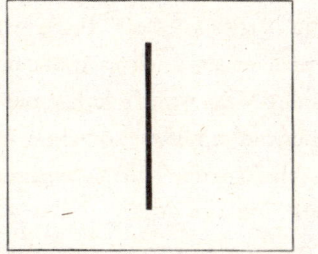 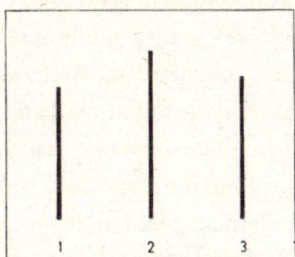

While the perceptual task was seemingly straightforward, a significant number of test subjects, when asked after participants who had been told to answer incorrectly, also themselves gave the wrong responses, choosing lines that were obviously either longer or shorter than the one being measured. They knew which answer was correct, but those around them were disagreeing. It was disconcerting, and for some the dissonance was overwhelming. As Asch later wrote, the lone subject of study "faced, possibly for the first time in his life, a situation in which a group unanimously contradicted the evidence of his senses." It was an undoubtedly harrowing and uncomfortable moment for the test subject, who was well aware of the correct answer but was seated next to seven individuals who were, often unanimously, making the wrong choice. For Asch, and many others, the fact that "reasonably intelligent and well-meaning young people" were "willing to call white black is a matter of concern," calling into question the educational systems that our culture had produced as well as our values as a society.

Asch's interest in conformity and the power of group pressure from a psychological perspective was a reflection of questions about human nature—about the human capacity to inflict harm on others—that had arisen in the wake of the rise of the Nazi Party in Germany in the 1930s. A friend and colleague would later recall that when it became clear that the number of "yielders" in his studies, as they labeled those who buckled under the pressure of the group, "was disappointingly large," they "all had to learn to swallow that result, along with the lessons of the Nazi successes." The experiments conducted by Asch, along with others such as those performed in the following years by Stanley Milgram at Yale, had put to rest any lingering hope that the American mind was somehow immune from the pressures of group psychology that had overwhelmed the German public across the Atlantic.

Asch's experiments marked the beginning of what some would describe as a golden age of social psychology in the postwar period. The institutional review boards that today carefully monitor proposed studies involving human subjects did not exist. The departments essentially were left to police themselves, and experiments on human subjects, including ones that required significant levels of deception, were frequently permitted at the time. While many would later challenge the ethics of allowing such experiments to proceed, given the extent of the deceit and manipulation involved, the tests arguably produced some of the most valuable research into social and group psychology that has ever been performed.

The obedience experiment conducted in 1961 by Milgram, who had studied under Asch at Princeton, went even further than the line comparison tests from a decade before. Milgram, who was an assistant psychology professor at Yale, designed his experiment on conformity in order to assess not merely whether test subjects would buckle under the pressure of a group when faced with a simple perceptual task, such as assessing the relative lengths of lines, but rather their willingness to inflict harm on innocent strangers when instructed to do so by an individual in a position of apparent authority. Milgram was born in 1933 in New York, and his father was a cake baker who had immigrated to the United States from Hungary. His mother had left Romania as a young child. Milgram's experiment involved the recruitment of hundreds of residents of New Haven,

Connecticut, to volunteer for what they had been told was a psychology experiment involving learning and punishment that was being conducted by Yale University. An advertisement seeking volunteers was placed in the local newspaper, and Milgram's team followed up by sending letters to randomly selected residents from the phone book to recruit additional participants. Each of the volunteers was paid $4, as well as fifty cents for taxi rides to and from the laboratory. The test subjects were told that in the experiment they would play the role of a "teacher," whose job would be to administer electric shocks to another individual, known as the "learner," in order to assess whether the shocks would assist the learner in memorizing random pairs of words, such as "blue" and "box," or "wild" and "duck," more accurately.

The electric shock machine was made to look authentic, and almost menacing, with knobs and lights, a buzzer, and various labels noting the level of voltage that would be administered by turning the knob to different positions. At the outset of each session, participants were even given a mild shock themselves from the machine in order to further convince the test subjects that they would be administering actual electrical voltage as part of the experiment. The learner, of course, was in on the ruse, and played by a forty-seven-year-old accountant. The electric shock machine emitted sounds and flashed lights, but could not harm anyone. As the amount of voltage increased throughout each session, the learner, however, would begin yelling and shouting, asking both the test subject and the experimenter to halt the experiment. The question was how far subjects would proceed notwithstanding his increasingly desperate pleas to stop. Of the dozens of individuals who participated in the experiment, a striking two-thirds complied with directions to administer what they had reasonably been led to believe was a harmful level of electrical voltage to an otherwise innocent test subject. The results captivated the country and sparked a debate about the human capacity for inflicting harm at the direction of authority figures.

In one of the most haunting sessions from the experiment, one of the volunteers, a fifty-year-old man, whom Milgram later described as "a rather ordinary fellow," at first protested mildly when asked to administer the series of increasingly strong shocks to the victim. As the voltage approached what appeared to be more dangerous levels, and the purported victim could be heard shouting repeatedly to be let free and to stop the experiment, the test subject attempted to dissuade the experimenter from asking that he proceed with the administration of a 180-volt shock.

> SUBJECT: I can't stand it. I'm not going to kill that man in there. You hear him hollering?
>
> EXPERIMENTER: As I told you before, the shocks may be painful, but—
>
> SUBJECT: But he's hollering. He can't stand it. What's going to happen to him?
>
> EXPERIMENTER (*his voice is patient, matter-of-fact*): The experiment requires that you continue, Teacher.... Whether the learner likes it or not, we must go on.

And go on he did. Over the next several minutes, the test subject proceeded to administer a series of escalating shocks through shouts of pain and protests from the victim, who pleaded repeatedly to let him out of the room and stop the experiment. The transcript of the exchange is absolutely striking. A certain decorum remained constant throughout the session, notwithstanding the fact that one man believed that he was shocking another to death. As Milgram put it, "A tone of courtesy and deference is meticulously maintained." For many, the dissonance between the measured dialogue of the test subject and the experimenter, on the one hand, and the cries of agony from the victim, on the other, challenged the view that the capacity to inflict harm on the innocent was solely the domain of the depraved. "He thinks he is killing someone," Milgram later wrote of the subject, "yet he uses the language of the tea table." We had collectively perhaps hoped that the destruction wrought during World War II had been the work of isolated actors, an aberration from the ordinary capacities of the human mind. Milgram's experiment provided a jarring and alternative explanation—that such a capacity was far more commonplace, and indeed banal, than we had ever considered.

Not all of Milgram's subjects, however, were as compliant. One woman, a medical technician from Germany who had grown up during the rise of the Nazi Party in the 1930s, stood out. At one point during her session, as the setting on the shock generator approached 210 volts, she paused, asking, "Shall I continue?" The investigator leading the session, who was a thirty-one-year-old biology teacher wearing a gray lab coat, replied, "The experiment requires that you go on until he has learned all the word pairs correctly." He also repeated that the shocks "may be painful" but were "not dangerous." The subject then escalated the interaction somewhat: "Well, I'm sorry, I think when shocks continue like this, they *are* dangerous. You ask him if he wants to get out. It's his free will." Her act of defiance almost seemed casual; its steely resolve both inspiring and unremarkable. Moments later, she told the experimenter that she would not proceed with shocking the victim at higher voltages and left. Milgram observed that "the woman's straightforward, courteous behavior" and "lack of tension" made her defiance appear to be "a simple and rational deed."

The psychological resilience that the woman displayed was what Milgram had expected would have been the case for most of those tested. His hopes, however, had been misplaced. Many of those who participated in the experiment proceeded to administer what they believed were significant doses of electricity to victims yelling for the experiment to stop. The prevalence and indeed ease with which so many submitted were, of course, stark reminders of our shortcomings as a species. But they also suggested a path forward, or at least exposed the psychological obstacles around which one must maneuver in business, in order to have any hope of creating something new.

* * *

The instinct toward obedience can be lethal to an attempt to construct a disruptive organization, from a political movement to an artistic school to a technology startup. At many of the most successful technology giants in Silicon Valley, there is a culture of what one might call constructive disobedience. The creative direction that an organization's most senior leaders provide is internalized but often reshaped, adjusted, and challenged by those charged with executing on their directives in order to produce something even more consequential. A certain antagonism within an organization is vital if it is to build something substantial. An outright dereliction of duty might simply hold an organization back. But the unquestioning implementation of orders from higher up is just as dangerous to an institution's long-term survival. The challenge for businesses is that executives and managers far too often select for and reward an unthinking compliance in those they hire—a simpleminded obedience that is corrosive to building a business capable of something more than execution on the whims of a founder.

The group of experiments by Asch, Milgram, and others—now classics in investigational social psychology—prompted an entire generation of psychologists and academics to question the ability of individuals to resist the pressure of authority and delivered something of a somber and enduring referendum on humanity. Some had hoped that the experience in Europe had been an aberration—that other nations, if tested, would not have succumbed and indeed submitted to totalitarian rule without more fierce resistance. As Howard Gruber, a psychology professor at Columbia University's Teachers

College who had studied under Asch, would later recall, the studies conducted by that era's researchers made clear "that conformity is international." America might have been exceptional, but not in all respects.

Some amount of what might be thought of as a sort of social deafness may, in this way, be productive in the context of building software. An unwillingness, or perhaps an inability, to conform to those around us, to the cues and norms put forth by others, can be an advantage in the realm of technology. A willingness to withdraw from the world, and to decline to engage with external views at certain critical moments of an organization's evolution, has been vital in the context of building Palantir over the past two decades.

Other purported disabilities have, in different domains, proven adaptive. In September 1922, Claude Monet, after months of declining vision, was diagnosed with a cataract, which, according to his Parisian eye doctor, had reduced the painter's vision "to one tenth in the left eye and to the perception of light with good projection in the right eye." He went through periods of seeing the world tinged by an orange hue, then weeks later a blue cast. A surgery, and the arrival of some German lenses, eventually helped address the problem. His later works grew increasingly and viscerally divorced from bare representation, including a canvas, inflected with hints of teal and crimson, titled *Weeping Willow*, whose "gestural lines," one art critic has noted, "blur the image until it veers into abstraction." A retrospective of the painter's work alongside that of the American artist Joan Mitchell, which opened in Paris in 2022, suggested that Monet was responsible for the rise to dominance of abstract expressionism that would follow in the decades after his death in 1926.

Similarly, when Ludwig van Beethoven began losing his hearing in his twenties, he was at first intensely guarded about his diminished capacity to listen to the very music that he was building a career composing. In 1801, Beethoven wrote to a violinist friend of his, "I beg you to treat what I have told you about my hearing as a great secret." As news, however, of his hearing loss became more widely known over the years, the public grew fascinated with his seemingly otherworldly capacity for musical composition "in spite of this affliction," as Beethoven's nephew wrote to his uncle. The question, of course, is whether the perceived disability was a disability at all—whether he was capable of composing such great works *in spite of* his incapacity or rather *because of* it. Some have argued that Beethoven's hearing loss merely redirected and perhaps augmented his creative process, forcing him to rely more heavily on the act of writing out his compositions, and thereby allowing him to construct "a novel sonic universe," as one music critic has written, "because he was being led by his eyes as much as by his memories of sound."

* * *

The instinct to conform to the behavior of those around us, to the norms that others demonstrate, and to prize the abilities that most around us find second nature, is in the vast majority of cases extraordinarily adaptive and helpful, for both our individual survival and that of the human species. Our desire to conform is immense and yet crippling when it comes to creative output. In Asch's experiments, there was a subset of those tested who reliably buckled under the pressure each and every time they were confronted with blatantly false reports of the relative lengths of the lines they were shown. Another cohort never wavered in correctly assessing their lengths notwithstanding the pressure of others. It is this insensitivity to a certain type of social calculation, and resistance to conformity, that has been vital to the rise of Silicon Valley's engineering culture.

Chapter Thirteen

Building a Better Rifle

On September 28, 2011, a group of twenty-four U.S. soldiers were on patrol in Helmand province in southern Afghanistan, supporting special forces personnel who were attempting to build relationships with village leaders in the region. The stretch of land in central Asia, at the precarious intersection of numerous empires over the course of three millennia, had been the subject of repeated cycles of invasion since at least Alexander the Great in the fourth century BC, who was himself ambushed and shot by an Afghan archer with an arrow during a campaign across the country from the Khyber Pass in the east to Persia in the west. On that September afternoon, the patrol stopped, and two marines got out of their vehicles to take a look around, likely searching for potential signs of roadside bombs that might have been hidden by Afghan insurgents along their route. Moments later, a bomb detonated, and the marines, wounded badly, fell to the ground. James Butz, a twenty-one-year-old army medic from Porter, Indiana, immediately rushed forward to help—not even sparing a moment to gather his own helmet and rifle. A second explosion then went off. "Two soldiers were down," his father later recalled. "Jimmy didn't hesitate." All three men, Butz as well as the two marines he was running to help, died that day.

The use of roadside bombs, which came to be known as improvised explosive devices, or IEDs, across Afghanistan against American and allied forces would expand significantly in the months that followed. By 2012, more than three thousand service members in the U.S. military had been killed by the handmade bombs that were hidden or buried beneath roads while insurgents waited out of sight to detonate them. A total of 14,500 IED attacks against U.S. and allied soldiers occurred across the country in 2012 alone. The bombs, whose explosive material was often made from widely available crop fertilizers, presented an escalating crisis for American forces, which had been sent to Afghanistan to build meaningful relationships and coalitions with local militias, in villages and towns that were scattered across the region—work that required constant travel and interaction with civilians. As a U.S. Navy officer who spent years searching for and defusing the bombs later observed, the IEDs forced U.S. soldiers to "confine themselves to massive, armored vehicles and travel at high rates of speed or plow through farmers' fields to avoid roads entirely."

The U.S. military spent more than $25 billion from 2006 to 2012 in an attempt to develop solutions to counter and defend against the crude explosive devices, which often cost less than $300 to make. The armored personnel carriers that ferried troops across Afghanistan were particularly vulnerable; their protective armor was simply too light to withstand the blasts given off by the roadside bombs that were hidden across the landscape. The U.S. Army decided to order a new fleet of vehicles with more substantial and protective ceramic composite material for armor. By October 2012, more than twenty-four thousand of the vehicles would be manufactured and sent to the battlefields of Afghanistan and Iraq. In response, however, insurgents simply began building bigger bombs, and ones that could be detonated remotely at greater and therefore safer distances. The more powerful explosive devices came to be known as buffalo killers by soldiers in the field for their ability to take out even the larger and more heavily armored vehicles that the military had ordered to respond to the threat.

BUILDING A BETTER RIFLE

By 2011, it had become clear to nearly everyone in the U.S. military that better intelligence was needed to assess the safety of particular roads and potential routes across the country, as well as to identify and capture the bomb makers themselves. The frustration of so many soldiers and intelligence officers in the field was that they had the information they needed—the records and locations of prior attacks, the types of bomb-making materials that had been used, the fingerprint scans and mobile phone numbers of captured insurgents, and the reports of confidential informants who had been recruited by American intelligence agencies, to name only a few of the data sets that were available. The information was sitting there, in dozens and hundreds of government systems, for anyone with the right clearance to access. The task of stitching it all together, however, into something useful—into something actionable that patrols could use as they planned their next route to visit a neighboring village, or decided which prisoners to question and what information they might provide—was often effectively impossible.

The structural issue was that those designing the army's software system at the time, including programmers at Lockheed Martin, in Bethesda, Maryland, were too far and too disconnected from the actual users of the software, the soldiers and intelligence analysts, in the field. The gulf between user and developer had grown too wide to sustain any sort of productive cycle of rapid iteration and development. The construction of any technology, including military software systems, requires an intimacy between builder and user—an emotional and often physical proximity that for many government contractors in the suburbs of Virginia and Maryland outside Washington, D.C., was as foreign as the Afghan insurgents American troops were fighting a world away. In another era, U.S. fighter pilots during World War II would frequently visit the factory of Grumman Corporation, the predecessor of Northrop Grumman in Bethpage, New York, on Long Island, to provide suggestions on the design and construction of the company's planes, including the F6F

Hellcat, which proved decisive in the air battle over the Pacific, according to the author Arthur L. Herman. In Afghanistan more than half a century later, however, that link between soldier and supplier had withered, if not been severed completely.

With the army's attempt to build a software system for soldiers in Afghanistan, the reliance on a tangle of contractors and subcontractors—and a yearslong procurement process that often involved more preparation and planning for the construction of software than actual coding—had deprived Lockheed Martin of any real opportunity to incorporate feedback from its users into its development plans for the system. The military's software project had devolved into a pursuit of an almost abstract conception of what software *should* look like, with far less concern for the actual features and capabilities, the workflows and interface, that would either make the software valuable to someone working all night on a laptop in Kandahar to prepare for a special forces operation the next morning or not.

An intelligence officer in Afghanistan with the 82nd Airborne filed a request in November 2011 with a relatively new division within the U.S. Army, the Rapid Equipping Force, which was located in Fort Belvoir, Virginia, just outside Washington, D.C., and had been established in 2002 as an attempt—one of dozens in recent decades—to expedite the development of new weapons, equipment, and software platforms for soldiers on the front lines. The organization's stated goal had been to acquire or build what soldiers needed within three to six months—a radically ambitious timeline in the world of defense contracting, where new weapons systems often languished in development for years and even decades. The intelligence officer submitted a formal request to the army procurement office in Virginia asking for access to Palantir's software to help gather and analyze intelligence from the field in Afghanistan in order to counter the growing threat from IEDs. The stakes were high, and growing. The officer wrote that the lack of access to Palantir's software had led to "operational opportunities missed and unnecessary risk to the force."

By early 2012, the requests for access to Palantir from soldiers in the field in Afghanistan had begun mounting, with some finding ways to circumvent the layers and bureaucracy of more traditional procurement channels in favor of sending their requests for laptops and software to senior military officers directly. In January 2012, for example, an intelligence officer in Afghanistan sent an email to army procurement personnel arguing that the army's data analysis system was "not making our job easier, while Palantir is giving us an intelligence edge." The following month, on February 25, 2012, the same officer repeated his request for Palantir, emphasizing the rising stakes of attempting to wage a war without effective software and the growing frustration from soldiers in the field. "We aren't going to sit here and struggle with an ineffective intel system while we're in the middle of a heavy fight taking casualties," the intelligence analyst wrote. A deputy to James Mattis, who would later become U.S. defense secretary, wrote in an internal request within the defense department for access to our software, according to an article in *Fortune*, "Marines are alive today because of the capability of this system."

For many even far from the battlefield, the thought of sending soldiers halfway around the world to fight a war, only to hesitate when those same soldiers are telling you they need better equipment in the field to stay alive, was absurd. The more fundamental issue was that a broader public disillusionment with American involvement in Afghanistan, as the years and casualties mounted, began to shape and warp discussions around what resources soldiers needed to do their jobs. We should, however, as a country, be capable of continuing a debate about the appropriateness of military action abroad while remaining unflinching in our commitment to those we have asked to step into harm's way. If a U.S. marine asks for a better rifle, we should build it. And the same goes for software.

An even more fundamental issue was that the political class setting the agenda in Afghanistan had itself never flown halfway around the world to risk one's life. Over twenty years, nearly 2,500 members of the U.S. military were killed in Afghanistan, in addition to approximately 70,000 civilians. The conflict would end up costing $2 trillion over two decades, or $300 million every day for twenty years, according to estimates by a research group at Brown University. It has been more than fifty years since the United States abandoned mandatory conscription in 1973, near the end of the Vietnam War. And since then a generation of political elites has essentially enlisted others to fight their wars abroad.

FIGURE 11

Percentage of Members of U.S. Congress
Who Have Served in the Military

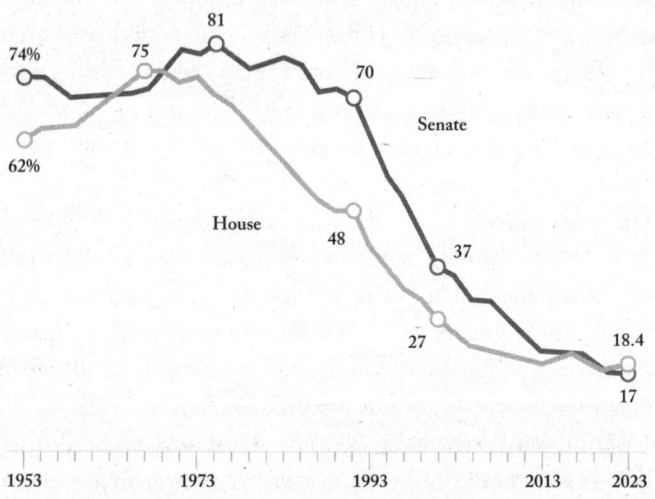

As of August 2006, there were only three members of Congress—three out of our 535 U.S. representatives and senators—who had a child serving in the American military. Charles Rangel, who represented New York City in Congress for nearly five decades from 1971 to 2017 and fought in Korea in the 1950s, has been a lonely proponent of reinstating the draft. He introduced legislation at least seven times in recent decades calling for the resurrection of conscription. If a battle abroad "is truly necessary," he has said, "we must all come together to support and defend our nation." The current model is utterly unsustainable. We should, as a society, seriously consider moving away from an all-volunteer force and only fight the next war if everyone shares in the risk and the cost.

A battle over which software intelligence platform to use in Afghanistan would continue for years. In the end, it was the individual soldiers and intelligence analysts who needed a better system, and the army's disinterest in adjusting more quickly when faced with criticism of its own incumbent platform, that began to shift the discussion. In the American system, imperfect as it may be, "you get things done by power," as Patrick Caddell, a political adviser to President Jimmy Carter, once said, and "you get power from having public support." The soldiers knew what they needed, and their voices would end up being heard. But it was also a little-known federal statute enacted in response to a prior conflict in another era and a different part of the world—a law that would essentially go overlooked for two decades—that helped tip the balance.

• • •

In the early 1990s, shortly after the U.S. military began its aerial bombardment of Iraq and sent troops to defend Kuwait, commanders in the U.S. Air Force identified an urgent, and seemingly unlikely, problem. The most powerful air force in the world, with the most advanced fighter jets ever produced and rocket-propelled missiles that could reach across continents, lacked something far more lowtech and less expensive. The U.S. Air Force personnel who had flooded into Kuwait in the wake of Saddam Hussein's invasion of the country did not have enough two-way radios, the handheld devices that were essential for fast communication across the new military bases that the United States was establishing. The radios, the kinds that are used on construction sites and camping trips, were readily available at stores across the country and could be purchased by anyone for less than $20 at a local electronics store.

The solution for the U.S. Air Force seemed simple: buy more. The best available model at the time was made by Motorola, the American electronics giant that was founded in 1928 in Schaumburg, Illinois. A Japanese subsidiary of the company had large quantities in stock of the radios that the air force needed, and an urgent order for thousands of them was placed. Motorola, however, hesitated when it received the request, which was accompanied by a long list of special provisions inserted by U.S. officials, including what the company believed were onerous and unnecessary requirements to produce data on the costs involved in manufacturing the radios. The litany of requirements was a standard part of the military procurement process at the time; its ostensible purpose was to ensure that the government received a fair price for what it bought. Motorola did not have anything to hide. The issue was that the company did not have the accounting systems in place that would have enabled it to track its manufacturing costs in the specific way that the U.S. government required. As a result, the company could not lawfully sell its radios to the American military.

The air force was in a bind. A war was mounting in Iraq, and the military did not have sufficient numbers of the most basic of tools—a working, portable communications device. The result seemed absurd. A patchwork of regulations that had been intended to *protect* the U.S. government against overspending were now preventing that same government from buying what it needed on the open market in the middle of the most significant military conflict in a generation. The air force contemplated attempting to navigate through its own regulations and find a work-around. But developing an alternative contractual model, one that would have avoided the cost disclosure requirements mandated by law, "would have taken some time," according to Lieutenant Colonel Brad Orton, who was leading the air force's effort to acquire the radios, "time that we didn't really have." In the end, Orton and others decided to circumvent the U.S. government's own regulatory regime entirely. They reached out to the Japanese government and arranged for Japan, not the United States, to purchase six thousand of the handheld radios directly from Motorola and then for the Japanese government to send them to the U.S. Air Force in Kuwait.

The episode came to symbolize the extent of internal dysfunction within the U.S. government procurement process, which had become so contorted and inefficient that the military, during wartime, was prevented from buying what any civilian could have purchased from a local electronics store. The challenge was systemic, and the roots of the dysfunction ran deep. Senator William Roth, who represented Delaware for three decades beginning in 1971, would later point out the absurdity of the fact that the federal government struggled to purchase products that anyone could "buy at the local Wal-Mart and Kmart."

The structural issue was that the procurement bureaucracy within the U.S. government had become so large and so entrenched, wielding enormous power and influence, that it had grown used to ordering custom-built versions of whatever it needed instead of shopping for goods, like everyone else, on the open market. The federal procurement officials responsible for supplying the U.S. military could direct the efforts of thousands of subcontractors and suppliers, essentially dictating that anything they wanted or needed be conjured and created from scratch. The government did not technically employ the product designers or own the factories. But it effectively controlled them, and could also pay any price. At the time, the U.S. government "tended to spend too much because it had almost everything it bought 'custom made' to government or military specifications,"

Al Gore, who worked on procurement reform during his time as vice president under Bill Clinton, wrote in 1998. The U.S. Army, for example, at one point in the 1990s drafted more than seven hundred pages of specifications on how to bake cookies, specifications that would be sent to its suppliers, instead of simply working with a major manufacturer whose cookies were already being made and on grocery store shelves.*

The roots of the problem, as well as the increasing public frustration with wasteful government spending, had been growing for nearly a century. A commission established by President Theodore Roosevelt in 1905, for example, had discovered that the U.S. government was purchasing 278 types of pens, 132 variations of pencils, and twenty-eight distinct colors of ink. Gifford Pinchot, a close friend of Roosevelt's who served on the commission, noted that the government had become "debased by generations of political control, sunk in the mire of traditional red tape"—a term that has its roots in the red-colored cloth tape that various governments, including that of the United States, had used throughout history to tie and bundle documents.

In the modern age, the constant rotation of personnel through the government, both in the military and in the civilian branches, incentivized inaction and complacency. In the early 1980s, a series of reports regarding the significant sums paid by the U.S. government for commonplace household items captured national attention, prompting calls for reform. In 1983, for example, the U.S. Navy reportedly paid $435 for an "ordinary hammer," according to a report in the *New York Times* at the time, and $400 for a "thumb-sized plastic knob" that was used in the cockpit of a fighter plane. Some of the prices that captured public attention were arguably misleadingly

---

* One list from the 1980s of military specifications for cookies mandated that the final product, baked pursuant to Sections 5.4.1.1 and 5.4.1.2 of the document, "shall yield tender and crisp cookies that are not peaked."

high. The hammers, for example, had been listed on an invoice as costing $435 each, even though that figure had been calculated by assigning a proportion of the labor and overhead involved in the production of more than four hundred other spare parts and pieces of equipment to each individual item delivered on an equal basis—an accounting method that imperfectly divided overhead costs across hundreds of items, including the hammers. Still, the public rightly sensed a system that had grown so large and unwieldy that it was nearly beyond reform, planting the seeds of discontent that have resurfaced today, nearly half a century later, about a Washington establishment focused solely on its own survival at the expense of the public interest and common sense. In 1984, a journalist described Joseph Sherick, the inspector general of the U.S. Department of Defense under President Ronald Reagan who had been charged with policing the federal procurement bureaucracy at the time, as an "alligator" patrolling "a 'swamp' of mismanagement and abuse at the Pentagon."

. . .

By the early 1990s, the reformers had essentially won the argument, and the public was ready, even eager, to see the size and scale of federal spending cut back. Bill Clinton, who won the presidency in 1992, had pitched himself to the American public as a pragmatic reformer—a Democrat who would trim government, not expand it. He would later say, in a State of the Union address during his first term, "We know there's not a program for every problem." Clinton cast himself as more closely aligned with skeptics of the federal bureaucracy, not its advocates. At a press conference in September 1993, announcing what he had described as a national performance review, which was intended to overhaul the federal bureaucracy, Clinton told reporters, "The Government is broken, and we intend to fix it." The country was receptive to the message, which had significant support across party lines. David E. Rosenbaum, a political correspondent for the *Times*, wrote the following day, "No one who has tried to fill out a Medicare claim form, get the Internal Revenue Service or the Social Security Administration on the telephone, apply for a Government contract—no one, in short, who has ever been hogtied by Federal red tape—can disagree with Mr. Clinton's description."

Clinton had been working with members of Congress on both sides of the aisle for months on a new federal statute aimed at reforming the federal procurement process. Shortly after 10:00 a.m. on October 26, 1993, Clinton gathered with his vice president, Al Gore, and others in the Old Executive Office Building at the White House to preview his planned reforms and announce a series of spending cuts to federal programs. The struggle by the air force during the Gulf War to purchase two-way radios from Motorola—and a furtive, last-minute deal with the Japanese government to avert a crisis—was, for Clinton, a clear example of why the U.S. Congress needed to move quickly to overhaul the system. "This should never happen again," Clinton said. Gore, who was standing by his side, added, "When the government of another nation has to step in and buy something for the U.S. military because our procurement regulations are so crazy, that's a clear wake up call."

The draft legislation that Clinton and others had planned was a bill that would grant the government far more discretion in purchasing decisions. The prevailing regulatory regime had focused on price and as a result often led to contracts being awarded to bids that offered the lowest cost irrespective of whether the contractors making them were best suited to do the job. The new legislation shifted the focus to value, as opposed to cost exclusively, providing the government with much broader discretion to make purchasing decisions that it believed were in the public interest. In addition, the bill introduced a new requirement that would essentially remain unused for more than two decades. The law, which would come to be known as the Federal Acquisition Streamlining Act of 1994, required that the government consider buying commercially available products, whether they were two-way radios or armored personnel carriers, before attempting to build something new from scratch.

The legislation attracted little attention at the time; it was the product of a sort of behind-the-scenes governance, without the promise of much publicity, that has lost favor in recent years. The bill was sponsored by John Glenn, the former astronaut and then senator from Ohio. His legacy was secure, and he had little to prove, to his constituents or to the world. Glenn was born in 1921 in Cambridge, Ohio, a small town on the edge of the Appalachian Mountains. He served in the U.S. Marines as a fighter pilot during World War II and later became one of America's earliest and most celebrated astronauts. By the time he began working on the Federal Acquisition Streamlining Act, Glenn was serving his fourth term as a U.S. senator. He was unencumbered by a need to prove something to the public, whose affection he had already secured.

At a Senate hearing on February 24, 1994, in which the draft legislation was discussed, Glenn made clear that the proposed law "certainly is not glamorous," but was rather concerned with what he described as the "'grunt work' of government, the stuff that makes government work day in and day out, and makes it work efficiently." Everyone knew that the existing system was broken. But real progress had proven elusive. As Glenn pointed out, "We have wrestled year-in and year-out with these same issues, and still have failed to enact any meaningful reform." The strategy of public servants, he added, was often "to just not make waves, to not disturb their careers, to not do anything unusual that might get them in trouble." And there are a lot of people who do not want to get in trouble. Steven Brill, the author and journalist who founded the *American Lawyer* in the late 1970s, has documented the striking scope of the federal procurement machine, which includes 207,000 federal employees who have been hired to manage government acquisitions and purchases. "The bloat is undeniable," Brill has written.

In October 1994, the Federal Acquisition Streamlining Act was signed into law. At the signing ceremony, Clinton joked that he was hesitant to approve the bill, for fear of depriving late-night comedians of fodder about government dysfunction. "What will Jay Leno do?" Clinton asked. "There will be no more $500 hammers, no more $600 toilet seats, no more $10 ashtrays." The new federal statute, originally codified in Section 2377 of Title 10 of the U.S. Code, required that the U.S. government, "to the maximum extent practicable," acquire "commercial items," when such products are readily available on the market, as opposed to attempting to build new products from scratch. The final language of the statute was broad and seemingly unobjectionable—so broad that some believed it would not amount to much. The law merely required that the federal government *consider* purchasing commercially available products before ordering or building something new. The stage was now set for a legal skirmish that would play out two decades later.

. . .

In Afghanistan, software made by Palantir had found a committed band of supporters, particularly in the U.S. Special Forces, with teams where intelligence, and the ability to quickly navigate across databases and stitch together context in advance of missions, were critical. But the army as a whole, with hundreds of thousands of active personnel scattered around the world, remained resistant to any sort of broader rollout of Palantir to the force. Its own software program, which the military had been building for more than a decade, was still under development. The Federal Acquisition Streamlining Act, more than twenty years from its passage, with its plain language requiring that federal agencies consider commercial products before building their own, seemed to present a path forward.

In 2016, Palantir filed a lawsuit in the U.S. Court of Federal Claims, in Washington, D.C., arguing that the army had refused to even consider commercially available alternatives to its own data and analytical platform. This sort of litigation was rare, if not nonexistent, because most government contractors were wise enough to avoid suing the government agencies they were hoping would become their customers. We saw things differently. A federal statute had simple, plain language requiring the army to at least consider buying software products that were on the market before attempting to build its own. The case came before Marian Blank Horn, who in November 2016 issued a 104-page ruling, concluding that "the Army failed to properly determine ... whether there are commercially available items suitable to meet the agency's needs for the procurement at issue," and that "the Army acted in an arbitrary and capricious manner" in failing to do so. In short, we had won.

In March 2018, the U.S. Army announced that it would be selecting one of two companies, Raytheon and Palantir, to develop its intelligence platform moving forward. John McCain, a former officer in the U.S. Navy and then U.S. senator from Arizona, wrote that it was the right decision, that after $3 billion of investment "it was time to find another way." A year later, in March 2019, the army announced that Palantir had won the entire contract. The U.S. military's turn toward the technology sector, and perhaps reluctant embrace of an insurgent startup to take over construction of the system, was, according to the *Washington Post*, "the first time the government had tapped a Silicon Valley software company, as opposed to a traditional military contractor, to lead a defense program of record." The shift marked a pivot by the U.S. Department of Defense toward software and technology, toward a sector that had repeatedly turned its own back on America and its military in favor of its focus on, and indeed seemingly boundless enthusiasm for, more easily monetized consumer offerings.

In 2011, while we were sending engineers to Kandahar and working on building a more capable analytical software platform for U.S. and allied intelligence agencies, the focus of Silicon Valley, with its own armies of venture capitalists and entrepreneurs, was far from the mountain passes and deserts of Afghanistan. Zynga, the video game maker that had built a following on the back of *FarmVille*, a social-networking game in which players competed to cultivate land and raise livestock, was the darling of the Valley at the time. In December 2011, the company went public at a valuation of $7 billion. The enthusiasm from Wall Street, and focus on monetizing the millions and billions of potential users and clicks for the taking, was palpable. "This is a revolution," a brokerage firm analyst told the *Times* on the eve of Zynga's IPO. Afghanistan, and the lonely and often deadly task of clearing dusty roads of hidden bombs, could not have felt farther away.

Zynga was anything but alone in its zeal for and interest in the consumer market. Groupon was another of the year's most watched IPOs, the darling of darlings with the venture community. The company provided discounts to consumers at local retailers. At a valuation of $25 billion, Groupon was set to become "the largest IPO by a venture-backed company in history," an article at the time in *Forbes* noted. The company, which is still in business, albeit barely, has plummeted since its IPO and is today valued at mere pennies for every dollar that it was once worth. The Zyngas and Groupons had the world's attention. Palantir, by contrast, was off on its own adventure, far from the consumer and, as a result, in the minds of many, the right path. Some employees thought we were foolish. Others left and went to work for this new generation of consumer startups. One early engineer quit because he didn't think our shares would ever be worth anything and wanted more cash compensation instead of equity in order to buy a high-end stereo. The market had spoken. And it was unfashionable to question its wisdom.

The technology sector had turned its back on the military, disinterested in wrangling with an overgrown bureaucracy and ambivalence, if not outright opposition, from the public at home. There were other, more lucrative consumer markets to conquer. It was, however, a tolerance and perhaps some degree of taste for conflict, and a stubborn pursuit of something, anything that worked—that engineering instinct—that gave Palantir a foothold.

Chapter Fourteen

A Cloud or a Clock

THE AMERICAN ARTIST Thomas Hart Benton, who painted murals in the early part of the twentieth century, declined to jettison his representational approach even as modernism seemed to be sweeping away forms of art that could be readily deciphered. He taught at the Art Students League of New York for years, and his most famous student, Jackson Pollock, seemed ambivalent about his teacher's influence; the two had a long tangle of a friendship. In an interview with *Art and Architecture* magazine in 1944, Pollock offered a bit of begrudging praise for his former instructor, explaining that "it was better to have worked with him than with a less resistant personality." Benton initially thought little of Pollock's canvases, describing them as "paint-spilling innovations" and "scorned the idea of their possessing any long-term value."

The modern enterprise is often too quick to avoid such friction. We have today privileged a kind of ease in corporate life, a culture of agreeableness that can move institutions away, not toward, creative output. The impulse—indeed rush—to smooth over any hint of conflict within businesses and government agencies is misguided, leaving many with the misimpression that a life of ease awaits and rewarding those whose principal desire is the approval of others. As the comedian John Mulaney has said, "Likability is a jail."

The casual and unrelenting pressure to revert to the mean, to do what has been done before, to eliminate the wrong types of risks from a business at precisely the wrong times, and to avoid confrontation is everywhere and often tempting. But the culture's move to accommodate the subjective reality of its students and employees has only inflamed the sense of grievance and affliction that some feel. The rise of trigger warnings and other forms of acquiescence behind which the left has zealously rallied for more than a decade has backfired spectacularly, by fostering a sense of harm that often does not exist. Richard Alan Friedman, a professor of clinical psychiatry at Weil Cornell Medical College, said in an interview that, beginning in 2016 or so, he began seeing an increase in reports of students alleging that they had been "harmed by things that were unfamiliar and uncomfortable," and that the language they used, describing unease upon hearing comments in class, for example, "seemed inflated relative to the actual harm that could be done."

This is a grievance industry, and it is at risk of depriving a generation of the fierceness and sense of proportion that are essential to becoming a full participant in this world. A certain psychological resilience and indeed indifference to the opinion of others are required if one is to have any hope of building something substantial and differentiated. The artist and the founder alike are often "the mad ones," as Jack Kerouac wrote in *On the Road*, "the ones who are mad to live, mad to talk, mad to be saved, desirous of everything at the same time." The challenge, of course, is that some of the most compelling and authentic nonconformists, the artists and iconoclasts, make for notoriously difficult colleagues.

In the context of a creative endeavor, such as a technology startup or an artistic movement, the blank slate of human desire poses a fundamental challenge. We instinctively look to one another for guidance as to what is desirable, and as a consequence the intentions of others are often adopted wholesale and without reflection, left to grow within ourselves. René Girard, the French anthropologist, observed the conflicts and rivalries between monkeys that arise when one member within a group selects a single banana out of many, all of which are identical. "There is nothing special about the disputed banana," Girard said in an interview in 1983, "except that the first to choose selected it, and this initial selection, however casual, triggered a chain reaction of mimetic desire that made that one banana seem preferable to all others."

Our earliest encounters with learning are through mimicry. But at some point, that mimicry becomes toxic to creativity. Some never make the transition from a sort of creative infancy. Much of what passes for innovation in Silicon Valley is, of course, something less— more an attempt to replicate what has worked or at least was perceived to have worked in the past. This mimicry can sometimes yield fruit. But more often than not it is derivative and retrograde. The best investors and founders are sensitive to this distinction and survive because they have actively resisted the urge to construct imperfect imitations of prior successes. The act of rebellion that involves building something from nothing—whether it is a poem from a blank page, a painting from a canvas, or software code on a screen—by definition requires a rejection of what has come before. It involves the bracing conclusion that something new is necessary. The hubris involved in the act of creation—that determination that all that has been produced to date, the sum product of humanity's output, is not precisely what ought or need be built at a given moment—is present within every founder or artist.*

For a startup, or any organization that seeks to challenge an incumbent, the sort of mindless conformity that dominates modern commerce—an unwillingness to risk the disapproval of the crowd—

---

* For Ernst Kris, the Austrian psychoanalyst, the creation of art involved two independent processes, the channeling of "impulses and drives," often sublimated and beyond the reach of expression, as well as "work," the "dedication and concentration" required for the elaboration of an idea. The first stage, he wrote in 1952, "is characterized by the feeling of being driven, the experience of rapture, and the conviction that an outside agent acts through the creator."

can be lethal. In 1841, Ralph Waldo Emerson published "Self-Reliance," his enduring broadside against religious dogmatism, in which he railed against individual weakness in the face of institutional pressure. "For nonconformity," he reminds us, "the world whips you with its displeasure." Emerson made clear that the desire to conform not merely to those around you but to one's prior views on a subject can be just as limiting and indeed hobbling. The permanence of our thoughts and writing on the internet for all time—and the zeal with which the crowd confronts individuals who dare to venture into public life with perceived inconsistencies in their prior statements—only risk confining us further, into a straitjacket of our former selves. But Emerson is right to ask, "Why drag about this corpse of your memory, lest you contradict somewhat you have stated in this or that public place? . . . Leave your theory, as Joseph his coat in the hand of the harlot, and flee." We count ourselves among those who have repeatedly fled, abandoning failed projects within days of a lack of progress being surfaced and deconstructing dysfunctional teams. At other times, we certainly have been more timid, proceeding far too cautiously to reverse prior judgments and investments, in both particular people and projects. But the public, investing and otherwise, is often far too unforgiving of retreats and pivots, of revisions to plans and missteps. Nothing of consequence is built in a straight line. A voracious pragmatism is needed, as well as a willingness to bend one's model of the world to the evidence at hand, not bend the evidence.

. . .

When Isaiah Berlin wrote his essay *The Hedgehog and the Fox*, in 1953, the computing revolution was still far off. But there is no question that the ferocity of Silicon Valley's ascent, and by extension that of the United States, stems in significant part from the culture of the small tract of land south of San Francisco, in which an almost ruthless pragmatism took hold. For Berlin, there was a "great chasm" between the hedgehogs among us in the world, "who relate everything to a single central vision, one system less or more coherent or articulate, in terms of which they understand, think and feel," and the foxes, "who pursue many ends, often unrelated and even contradictory, connected, if at all, only in some de facto way." Berlin built something rich and enduring upon the thinnest of foundations— a single line, a fragment of a poem from the Greek poet Archilochus, who was born on an island in the middle of the Aegean Sea in the early seventh century BC. "The fox knows many things," Archilochus wrote, "but the hedgehog knows one big thing." And Silicon Valley is the consummate fox.

The founders and technologists who have constructed and will continue to construct the modern world willingly abandoned grand theories and overarching belief structures to build, indeed often build anything, as long as it worked. The distinguishing feature of technology, and in particular software, is that either it runs or it does not. There is no halfway, no *almost*, when it comes to software. The programmer is confronted with failure immediately. No amount of discussion or posturing can change whether the program performed as it should. Herbert Hoover, who studied geology at Stanford University, worked in the mining industry for nearly two decades, first during the gold rush in the 1890s in Western Australia, then a British colony, and later in Tianjin, China. He wrote in his memoirs that the "great liability of the engineer compared to men of other professions is that his works are out in the open where all can see them," and that the engineer "cannot bury his mistakes in the grave like the doctors," or "argue them into thin air or blame the judge like the lawyers." It is this sensitivity to results, and to failure, and perhaps an abandonment of grand theories of how the world ought to be, or how things ought to work, that is the seed of an engineering culture.

It is essential that the engineer—whether of the mechanical

world, the digital, or even perhaps the written—descend from his or her tower of theory into the morass of actual details as they exist, not as they have been theorized to be. One must, as the American philosopher John Dewey wrote in his essay "Pragmatic America" in 1922, "get down from noble aloofness into the muddy stream of concrete things."* An emotional and often physical proximity to the mess of imperfections and apparent contradictions of the systems and processes that one is charged with shaping is the source of progress, not its impediment. A commitment to this sort of pragmatism, or indeed the engineering mindset that has given rise to the Valley, "discourages dogmatism," as Dewey wrote, "arouses and heartens an experimental spirit which wants to know how systems and theories work before giving complete adhesion," and "militates against too sweeping and easy generalizations."

A certain ravenous pragmatism and insensitivity to calculation had been lost on the current generation. After the end of World War II, U.S. defense and intelligence agencies launched a massive and secret effort to recruit Nazi scientists, in order to retain an advantage in the coming years in developing rockets and jet engines. At least sixteen hundred German scientists and their families were relocated to the United States. Some were skeptical about this late embrace of the former enemy. An officer in the U.S. Air Force urged his commander to set aside any distaste for recruiting the German scientists to this new cause, writing in a letter that there was an immense amount to be learned from this "German-born information," if only "we are not too proud."

* * *

---

* Dewey took pride in the fact that pragmatism, as he wrote, "was born upon American soil."

In his book *Expert Political Judgment*, published in 2005, Philip E. Tetlock, a professor of psychology at the University of Pennsylvania, recounted being shown a demonstration in the 1970s that "pitted the predictive abilities of a classroom of Yale undergraduates against those of a single Norwegian rat." The challenge was to determine on which side of a small maze, left or right, a piece of food would be hidden. The experimenters would place food on the left side of the maze 60 percent of the time and the right side 40 percent of the time using a randomized selection process. The Yale students watched the rat attempt to ferret out the food, puzzling over potential patterns and grander schemes that might have lurked behind its placement. The rat, however, simply wanted to eat. And it turns out, the rat, not the undergraduates, was better at predicting where the food would be.

As Tetlock explained, the human mind was bested by the animal in the maze study "because we are, deep down, deterministic thinkers with an aversion to probabilistic strategies that accept the inevitability of error." The search for grand theories, for underlying systems and mechanisms of action in the world, in any other number of domains, from physics to medicine, has provided us with an enormous advantage, Tetlock acknowledged. Eugene Wigner, a theoretical physicist who was born in Budapest in 1902, famously observed the "uncanny usefulness of mathematical concepts." But that same drive for systematic theories of the world, for coherence at the expense of an effective muddle, has also left us with a persistent blind spot and resistance to embracing the instruction that the universe provides, even if its internal logic may be beyond us.

Tetlock's broader interest and project involved testing the accuracy of predictions made by political experts when confronted with questions about developments in global affairs. He and his team solicited and compiled a total of 27,451 forecasts made by experts starting in the 1980s, covering a range of political questions from the fate of the Soviet Union, whether South Africa would continue to maintain minority rule, and if Quebec would secede from Canada. Tetlock was interested in assessing which experts, among his panel, would be able to "'beat' the dart-throwing chimp" in making predictions about future historical events. It turns out that the 284 experts, that is, the academics and policy wonks selected to participate in Tetlock's study over the course of nearly two decades, did not generally fare better than chance. Some of the nearly three hundred experts, however, did outperform.

Tetlock had divided his specialists into groups of thinkers—foxes and hedgehogs—based on their responses to survey questions regarding the way that they approached intellectual challenges and problem-solving. And the foxes won.

FIGURE 12

Accuracy of Predictions Made by "Foxes" and "Hedgehogs" in Philip Tetlock's Review of 284 Experts

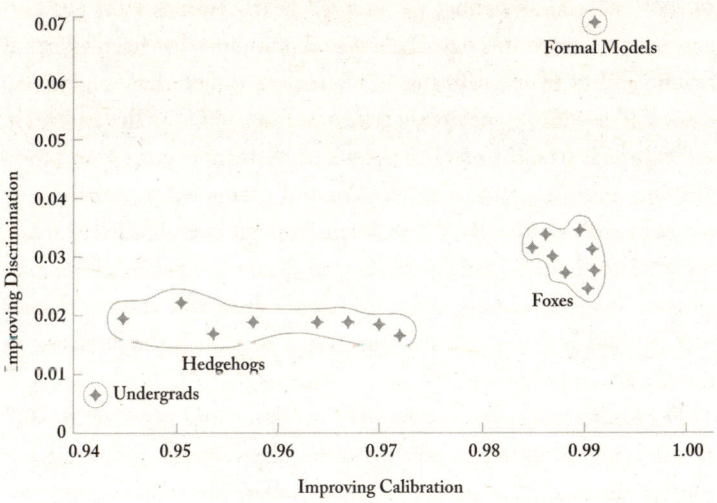

There are a number of ways to measure what Tetlock described as "foxiness." One could simply ask the expert whether he or she identifies more as a fox or a hedgehog, while explaining Isaiah Berlin's framework. And Tetlock did. But he also posed other questions to the experts, including whether they believed politics was more "cloudlike" or "clocklike," in an effort to tease out some of the same types of instincts. Those who described politics and history as more like a cloud than a clock, with its mechanistic precision and regularity, turned out to be significantly better predictors. The "worst performers," according to Tetlock, "were hedgehog extremists making long-term predictions in their domains of expertise."

* * *

In the late 1970s, Taiichi Ohno, a senior executive at Toyota Motor Corporation, published a book describing the Japanese automaker's reinvention of industrial manufacturing and articulated an approach to root-cause analysis that we adopted nearly twenty years ago and continue to use to this day. The method of inquiry has been essential in our ability to identify the fundamental, rather than superficial, causes of issues that inevitably arise across a company. The approach, on its face, is straightforward: ask why a problem occurred, and then ask why again four more times. We and others call it, very inventively, of course, the Five Whys. In the context of an industrial manufacturing facility, Ohno provided an example of a machine that stopped working because of an overloaded fuse, which upon further inquiry had been caused by a broken pump and ultimately worn metal parts.

For Ohno, who was born in 1912 in Manchuria just after the fall of the Qing dynasty, the method of inquiry focused on identifying the engineering flaws at the root of a problem. His father worked for the South Manchuria Railway, which was operated across an outpost of the Japanese empire in northeast China. Identifying the reasons for the failure of a system, whether it be an enterprise software platform or an assembly line for internal combustion engines, necessarily requires a focus on the inner workings and mechanics of the system at issue.

At Palantir, we build on this method of inquiry to incorporate an analysis and indeed acknowledgment of the human systems that are precursors to the software that we are building. Why did an essential update to an enterprise software platform not ship by a Friday deadline? Because the team had only two days to review the draft code. Why did the team have only two days to review? Because it had lost six software engineers in the budget review cycle late last year. Why did its budget decrease? Because the head of the group had shifted priorities elsewhere at the request of another group lead. Why was the request made to shift priorities? Because a new compensation model had been rolled out incentivizing growth in certain areas over others. And one can go even further, of course. Why were certain areas selected at the expense of others? Because of an ongoing feud at the company between two senior executives.

In this example, a missed deadline for shipping an update to a software system was, at its root, caused not by an individual engineer's oversight or even the team's failure to think ahead, but rather by an ongoing and increasingly adversarial interpersonal conflict at the highest rungs of the company. This sort of corporate butterfly effect is anything but new to those whose professions require subjecting oneself and submitting to the vicissitudes of modern corporate life. But what we have found is that those who are willing to chase the causal thread, and really follow it where it leads, can often unravel the knots that hold organizations back. It takes persistence and a willingness to dig beyond the first layers of a problem. The psychological dispositions and decision-making instincts of leaders within the company are often at the core of the challenge.

The exercise works most effectively if those involved resist the urge to assign blame to their colleagues and instead focus on the structural—and indeed often interpersonal—issues that gave rise to the mistakes at hand. We have conducted thousands of these Five Whys reviews over the past twenty years and draft detailed written reports that attempt to document, without assigning blame to individuals, the systemic and root causes of the problems that arise. The reasons for any complex system's failure, human and otherwise, can often feel beyond reach because of the difficulty, and patience required, of tracing the multiple and related chains of causation that lead through the labyrinth of the institutions and incentives we construct. A mistake, such as a missed deadline or lackluster product launch, often finds its root in the tangle of human relationships that make up the organization involved in the endeavor. The approach is an outgrowth of an engineering culture that at its best is unwaveringly focused on understanding what is working well and what is not. The challenge is fostering a sufficiently gentle and forgiving internal culture that encourages the most talented and high-integrity minds within an organization to come forward and report problems rather than hide them. Most companies are populated with people so fearful of losing their jobs that any hint of dysfunction is quickly covered up. Others are simply trying to make it to their retirement without being discovered as providing little or no value to the organization. Many more are monetizing the decline of empires they had once built.

It is a willingness to respond to the world as it is, not as we wish it might be, that has been a principal reason that the latest generation of Silicon Valley behemoths have come as far as they have. As Lucian Freud, the German-born figurative painter, perhaps the most enduring of the twentieth century, put it, "I try to paint what is actually there." The act of observation, of looking closely while suspending judgment—taking the facts in and resisting the urge to impose one's view on them—sits at the heart of any engineering culture, including ours. Freud, who was born in Berlin in 1922, was the grandson of Sigmund Freud, the psychoanalyst whose interrogations of the human mind transformed our willingness and ability to investigate our own psychology. The act of penetrating observation was essential to Lucian's portraits, which he has described as a sort of negotiation between artist and subject. They are unsparing and quite intimate, both bracing and gentle. His gaze, long and patient, sits at the heart of his work. Martin Gayford, a British art critic, has said that Freud "revived the figurative tradition" in the last century, a tradition that had fallen out of favor and was at risk of being eclipsed entirely. The artist once told an interviewer, "It can be extraordinary how much you can learn from someone, and perhaps about yourself, by looking very carefully at them, without judgment." It is this approach to observation, to looking closely at the clouds around us, while suspending judgment, that forms the foundation of the engineering mindset. The challenge we now face, in rebuilding a technological republic, is directing that engineering instinct, an indeed ruthless pragmatism, toward the nation's shared goals, which can be identified only if we take the risk of defining who we are or aspire to be.

Part IV

Rebuilding the Technological Republic

Chapter Fifteen

Into the Desert

IN LATE 1906, FRANCIS GALTON, a British anthropologist, traveled to Plymouth, England, in the country's southwest, where he attended a livestock fair. His interest was not in purchasing the poultry or cattle that were available for sale at the market but in studying the ability of large groups of individuals to correctly make estimates. Nearly eight hundred visitors at the market had written down estimates of the weight of a particular ox that was for sale. Each person had to pay six pennies for a chance to submit their guess and win a prize, which deterred, in Galton's words, "practical joking" that might muddy the results of the experiment. The median estimate of the 787 guesses that Galton received was 1,207 pounds, which turned out to be within 0.8 percent of the correct answer of 1,198 pounds. It was a striking result and would prompt more than a century of research and debate about the wisdom of crowds and their ability to more accurately make estimates, and indeed predictions, than a chosen few. For Galton, the experiment pointed to "the trustworthiness of a democratic judgment."

But why must we always defer to the wisdom of the crowd when it comes to allocating scarce capital in a market economy? We seem to have unintentionally deprived ourselves of the opportunity to engage in a critical discussion about the businesses and endeavors that ought to exist, not merely the ventures that could. The wisdom of the crowd at the height of the rise of Zynga and Groupon in 2011 made its verdict clear: these were winners that merited further investment. Tens of billions of dollars were wagered on their continued ascent. But there was no forum or platform or meaningful opportunity for anyone to question whether our society's scarce resources *ought* to be diverted to the construction of online games or a more effective aggregator of coupons and discounts. The market had spoken, so it must be so.

We have, as Michael Sandel of Harvard has argued, been so eager "to banish notions of the good life from public discourse," to require that "citizens leave their moral and spiritual convictions behind when they enter the public square," that the void left behind has been filled in large part by the logic of the market—what Sandel has described as "market triumphalism." And the leaders of Silicon Valley have for the most part been content to submit to this wisdom of the market, allowing its logic and values to supplant their own. It is our own temerity and unwillingness to risk the scorn of the crowd that have deprived us of the opportunity to discuss in any meaningful way what the world that we inhabit should be and what companies should exist. The prevailing agnosticism of the modern era, the reluctance to advance a substantive view about cultural value, or lack thereof, for fear of alienating anyone, has paved the way for the market to fill the gap.

The drift of the technology world to the concerns of the consumer both reflected and helped reinforce a certain technological escapism—the instinct by Silicon Valley to steer away from the most important problems we face as a society toward what are essentially the minor and trivial yet solvable inconveniences of everyday consumer life, from online shopping to food delivery. An entire swath of challenges from national defense to violent crime, education reform to medical research, appeared to many to be too intractable, too thorny, and too politically fraught to address in any real way. Most were content to set the hard problems aside. Toys, by contrast, did not talk back, hold press conferences, or fund pressure groups. The tragedy is that it has often been far easier and more lucrative for Silicon Valley to serve the consumer than the public, and certainly less risky.

• • •

The question of whether science and technology should be deployed to address violent crime in the United States has always been provocative. The history of abuses of power by U.S. law enforcement agencies, including by the FBI under J. Edgar Hoover and others, and incursions into the private lives of American citizens, is beyond dispute. An FBI file on the writer James Baldwin had swelled to 1,884 pages by 1974. Such invasions of personal privacy set the stage for a certain dualism in the debate over the twentieth century; either technological advances, including fingerprints, DNA, and later facial recognition systems, were essential to the difficult and often fruitless task of dismantling violent criminal networks, or they were the tools by which an overreaching state would target the powerless and imprison the innocent.

The next wave of technical breakthroughs, including the deployment of artificial intelligence to assist police departments, will only fuel this debate further and is set to reshape our sense of the possible when it comes to law enforcement and computing. A number of defense contractors, for example, including BAE Systems, working with the National Physical Laboratory in the United Kingdom, have developed gait recognition systems—software programs that are capable of identifying an individual based on little more than video footage of the person walking, without any access to an image of the individual's face. The technology has been under development for more than a decade and is improving in accuracy every day. Small flying drones operated by police departments can now approach a car window and break the glass, allowing police officers to take an unobstructed shot at someone within.

Our fear, of course, is that these sorts of emerging technologies might be used and misused, intentionally or otherwise, to detain or harm the innocent. The possibility of even a single abuse of the software that we are building must guide its construction and deployment. The administration of criminal justice is not the place for pragmatism, for some permissible degree of tolerance for error. François-Marie Arouet, better known by his pen name, Voltaire, wrote in 1749 that it would be preferable to set two guilty men free rather than imprison one who is "virtuous and innocent." In the eighteenth century, William Blackstone, one of England's greatest legal minds, went further, writing that it would be better to allow "ten guilty persons escape than that one innocent suffer"—a ratio that would come to structure debate about errors, permissible or otherwise, in criminal justice. Thomas Starkie, a British academic and lawyer who was born in the late eighteenth century, argued for allowing ninety-nine guilty criminals or more to walk free in order to ensure that a single innocent person would not be wrongfully imprisoned. The problem is not a fulsome and contentious debate about the merits of incorporating new technologies in the context of policing or criminal investigations. Rather, a fear of the unknown is too often used to abdicate responsibility for navigating any degree of uncertainty or complexity, and indeed possibility that technology could be misused.

Attempts to deploy software alongside law enforcement agencies in American cities have continued to be met with significant skepticism and distrust. In 2012, Palantir began working with the New Orleans Police Department to provide officers with access to the same software platform that had been used by U.S. Special Forces and intelligence analysts in Afghanistan to predict the placement of roadside bombs and capture those making them. The challenge for police officers in New Orleans and across the country was similar to what the U.S. Army had faced in attempting to disrupt the proliferation of bombs that were killing soldiers: too much information, and a complete lack of the underlying software architecture that would allow such information to be integrated and analyzed in any meaningful way. Criminal investigators and police officers in New Orleans needed a better system for stitching together the patchwork of information they had about criminal networks and tackling gun violence. The use of our platform, known as Gotham, spread quickly across the police department, with the *Times-Picayune* describing the system as "a one-stop shop for pulling up and cross-referencing information," and "discovering unseen connections among victims, suspects or witnesses."

The critics, however, were swift and fierce. The reaction, indeed, was visceral for many. Why should New Orleans permit the deployment of a software system designed for use in a foreign war on the streets of the city at home? In an essay published in 2018, a policy analyst with the American Civil Liberties Union wrote that the use of data in the context of law enforcement was "deeply problematic," given the threats to the civil rights and liberties of individuals who might be unfairly and unconstitutionally targeted by law enforcement as a result of the use of analytical software by the police. The moral outrage and indignation were directed against the application of a novel technology instead of the failure of the city's government to guard its residents. The country spent $25 billion to protect soldiers in Afghanistan from the threat of roadside bombs, but when it came to preventing the loss of American lives in our nation's cities, at the hands of the depraved, the mentally ill, and often extraordinarily well-resourced and ruthless violent gangs, the collective reaction is more often one of apathy and resignation.

Other technology firms have attempted, and abandoned, similar projects involving the use of software and artificial intelligence in the context of local law enforcement. In June 2020, Amazon decided to prohibit the use of its widely available and popular facial recognition software by police departments, after the company faced criticism that its system might be used to wrongfully target the innocent. That same month, IBM went even further, announcing that it would abandon all research and development into facial recognition capabilities. The company's chief executive officer sent a letter to Senators Cory Booker and Kamala Harris, among others, expressing his company's opposition to the use of the technology "for mass surveillance, racial profiling," and "violations of basic human rights and freedoms." The letter was representative of an ascendant form of hollow and meaningless corporate pronouncement, condemning an evil for which nobody is advocating. The subtle, interesting, and difficult discussion was not whether the abuse of such systems was justified but rather whether their proper use had any role to play in stemming violence in our cities. Thousands of people are murdered every year in this country. Hundreds of thousands and arguably millions more live in the shadow of such violence. For many critics of the use of software by local law enforcement, those lives hardly seemed to matter much in the moral calculus.

The rest of the country, and many politicians across the United States, have essentially shrugged when it comes to violent crime, abandoning any serious efforts to address the problem or take on any risk with their constituencies or donors in coming up with novel solutions and experiments in what should be a desperate bid to save lives. The price imposed on entrants into these areas has become incredibly high. And the message, implicit and often explicit, to those in Silicon Valley and across the technology sector has been plain. Steer clear. It was a deeply cynical response to violence that many of those in power in the United States have essentially abandoned any responsibility for addressing. Our representatives in Washington and elsewhere have simply turned their attention to less controversial terrain. Vast swaths of the American landscape, from law enforcement to medicine to education, have become innovation deserts where the Valley has been told, and often warned repeatedly, not to tread.

* * *

The view that advanced technology and software have no place in local law enforcement is an archetypal "luxury belief," to use the term of the author Rob Henderson. Such beliefs are ones that a privileged elite can afford to take on, almost as a cloak, as the columnist David Brooks of the *New York Times* put it, but that strike many as woefully "out of touch to people in less privileged parts of society." For those living under the constant assault of gunfire, for example, the thought of reducing support and funding for law enforcement struck many as an odd joke, the sort of campaign that had more to do with advancing a perception of political victory than actually shaping or advancing any outcomes on the ground.

The more fundamental issue is that the left establishment has decided, essentially unilaterally, that it need not be in conversation or dialogue with the right—that mere engagement with the other is itself a sign of cultural betrayal. When Peggy Noonan noted in a 2019 essay that the distaste by the Washington establishment for the current brand of American populism was, at its core, "almost aesthetic," she was absolutely correct in identifying the left's most pernicious weapon: the ability to brand an entire swath of political views—on issues ranging from national security, immigration, abortion, to law enforcement—as essentially lowbrow and uncouth. This is where Silicon Valley and other progressives have unfortunately and unwittingly deprived themselves of power in the cultural conversation. Their refusal to engage with the political claims and demands of essentially half of the country risks marginalizing their own agenda.

We have begun to privilege the symbolism of victory, the more theatrical elements and outward displays that constitute expression of our own moral superiority, over actual, and often less than visible, advances and improvements in standards of living and quality of life. And yet it is the zealous pursuit of those advances and outcomes that forms the bedrock of the engineer's approach to the world and the basis of a technological republic. The risk is that we abandon a moral or ethical system oriented around results—the outcomes that matter most to people (less hunger, crime, and disease)—in favor of a far more performative discourse, where the management of messages around such outcomes eclipses the importance of the outcomes themselves. And the reconstruction of a technological republic will, among other things, require the rebuilding of an ownership society, a founder culture that came from tech but has the potential to reshape government, where nobody is entrusted with leadership who does not have a stake in their own success.

Chapter Sixteen

Piety and Its Price

IN FEBRUARY 2023, the Economic Club of Washington, D.C., held a talk with David Rubenstein, the famed private equity investor, and Jerome Powell, the chairman of the Federal Reserve. The discussion covered familiar and expected terrain, including the debate about inflation and the appropriate level of interest rates, before taking an unexpected turn. At one point, Rubenstein, co-founder of the Carlyle Group with an estimated net worth of nearly $4 billion, asked Powell a seemingly straightforward question: "What is the salary of the chairman of the Federal Reserve Board?" Powell smiled, barely betraying even the slightest discomfort, and responded that his annual salary was roughly $190,000. Rubenstein then ventured further, asking Powell, "You think that's a fair salary for the job?" Powell replied, earnestly and plausibly, "I do." The audience laughed nervously, perhaps out of solidarity with Powell, who was handling a potentially volatile line of questioning with extraordinary grace.

It was a surreal moment. One billionaire asking a multimillionaire whether a salary of less than what a first-year associate would make at an investment bank was appropriate for the chairman of the Federal Reserve, the most powerful and influential central bank on the planet. The decisions that Powell himself makes are easily some of the most consequential in the world. During the course of his tenure, the fates of hundreds of millions of workers in the United States and abroad have hinged on his instincts about the path of inflation, the timing of interest rate increases and potential decreases, and his views about the strength of the American and global economies. Trillions of dollars in stock markets from New York to London, and Sydney to Shanghai, would trade hands as the direct result of his thinking and attempt to steer the U.S. economy, and by extension the world's, through a historically vulnerable period of inflation and potentially softening growth. And yet Congress has decided to pay him around $190,000 per year. In the private sector, such a salary would be considered absurd, given the scale and impact of the role and the resources available to his employer.

At that salary, Powell is essentially volunteering his time to the country. His compensation as an employee of the federal government is negligible with respect to his net worth, which has been reported to be in excess of $20 million, and he has said publicly that he essentially lives off his significant savings. But why are we, as a country, the world's wealthiest, asking for a volunteer to run the Federal Reserve? What incentives does that create, and how dramatically does that winnow the pool of potential candidates who might be interested in the job?

We complain about the influence of money in politics, only to remain silent as wealthy individuals increasingly dominate political races. The unintended consequence of our approach to public sector compensation is that an increasingly disproportionate number of the world's wealthiest are running for and winning public office, both in the United States and around the world. Of two thousand individuals identified as billionaires by *Forbes*, for example, a group of researchers at Northwestern University concluded in a 2023 study that approximately 11 percent of them had either held or run for political office. The incentives that our current approach creates are perverse. Members of the U.S. House of Representatives and the U.S. Senate earn $174,000 per year on average, even as their decisions have the potential to affect the lives of millions of soldiers, teachers, workers, and students across the country. Any business that compensated its employees in the way that the federal government compensates public servants would struggle to survive.

We tell ourselves that politicians should seek office for more noble reasons, those other than remuneration, only to pay them a fraction of what some of them could earn in the private sector. But we decline to confront the consequence of this approach, which is that we essentially incentivize candidates for public service to become wealthy before entering office, or to monetize their position after their departure. The extent of self-promotion and theater in the U.S. Congress is astounding, with representatives in the lower chamber vying for clicks and social media influence, and by extension incomes, after they leave office. The quality of candidates is a feature, in part, of what we are willing to pay them.

Others have advocated for increasing the pay of our elected and unelected representatives. As Matthew Yglesias, who co-founded *Vox* in 2014, has written, "If we want a better, more functional Congress, the American people should do what any other employer would do: make the job more desirable so that a larger pool of people run for office." In recent decades, numerous proposals have been made to reform public sector compensation in the United States, and most have gone nowhere. Since the founding of the republic, we have sought to hold on to the hope that well-meaning and talented people would run for office to serve their country for reasons other than their personal enrichment. In 1787, at a debate regarding congressional salaries, James Madison, who would become the fourth president of the United States, was skeptical of allowing members of Congress to have control over their own compensation. He argued that it would "be indecent to put their hands into the public purse for the sake of their own pockets." Yet our reluctance to blend personal incentives and public purpose, to adapt the practices of the business sector when setting salaries and compensation structures for government officials, will only hold us back. More experimentation, not less, is needed. And a far more radical approach to rewarding those who create the value from which we all benefit will be required.

In November 1994, Lee Kuan Yew, who served as the first prime minister of Singapore, was caught in a debate with other members of parliament regarding his proposed increases to government salaries. Lee had instituted a system under which the compensation of the island nation's public officials was set based on comparable salaries in private sector professions, including banking and law. By 2007, for example, the average annual salary of the country's ministers would rise to $1.26 million per year. Lee's critics argued that increasing salaries would attract the wrong type of candidate, those motivated to pursue government work for personal gain as opposed to public service. At a parliamentary debate on the matter, Lee responded that politicians "are real men and women, just like you and me, with real families who have real aspirations in life." He continued: "So when we talk of all these high-falutin, noble, lofty causes, remember at the end of the day, very few people become priests."

It is a skepticism of incentives in the domains that are most important to our collective good that may be part of what is holding us back. Why should we, the public, cede the use of incentives to the finance and banking industries, as well as the technology sector? The ascetic streak in American culture is admirable; deprivation, a skepticism of the material, reminds us that a bare and hollow commitment to consumption alone will inevitably lead us astray. But those instincts, the unstated desire that public servants be our priests, are having the unintended and undesirable consequence of depriving vast sectors of the public economies—in government, education, and medicine—of the benefits that the right incentives can create. Our reluctance to experiment with novel compensation models in the context of public pursuits is also deeply regressive, walling off entire professions—across the arts, medicine, government, publishing, and academia—as essentially the domain of an educated and often hereditary elite who can afford to volunteer its time and labor to the republic. A more uncharitable telling of the story would be that such elites do not want the competition to the high-status professions over which they currently enjoy near-exclusive access. We must pay our doctors and public servants and teachers more. These are noble callings. But those who pursue them should not be asked to accept their nobility as payment.

* * *

On the evening of May 31, 1953, in a remote area of eastern Idaho, a group of engineers from the U.S. Navy gathered to test the operation of a small nuclear reactor, one that would go on to change the balance of power over the world's oceans for the next half century. The distinction of this particular reactor was that it could fit on board a submarine, and the plan, radical at the time, was to have it power the ship. Experiments to reliably control and harness the power of nuclear chain reactions were still in their infancy, and the risks of an accident—including the leakage of radiation or an uncontrolled explosion—were significant. Everyone present "knew the danger," Edwin E. Kintner, the naval officer supervising the test, recalled years later. The hope was that the reactor could power a steam turbine; the fear was that it would turn into a nuclear bomb.

On that evening in the Idaho desert, Thomas E. Murray, the commissioner of the U.S. Atomic Energy Commission, pressed his hand to engage the reactor's throttle, and steam began spinning the heavy turbine. The nuclear engine, the first of its kind, ran for nearly two hours. The following month, the same reactor would be tested for five days straight. A race had begun, pitting the United States against the Soviet Union, to develop the next generation of submarines, ones that could maneuver through the oceans undetected—with a whisper rather than the drone of a diesel engine—and without the need to refuel.

The reactor worked, nearly flawlessly. In May 1955, the world's first nuclear-powered submarine, named the USS *Nautilus* after the craft in Jules Verne's *Twenty Thousand Leagues Under the Sea*, set off from New London, Connecticut, for San Juan, Puerto Rico, remaining submerged for nearly four days straight over the thirteen-hundred-mile journey. A U.S. Navy report would later note that the vessel was "almost immune to air attack" or detection, and could, with its speed, even evade a conventional torpedo. America was now positioned to retain an advantage over the oceans that would endure for decades, one which an adversary has yet to seriously challenge.

The plan to construct a sufficiently small nuclear reactor capable of powering a submarine had been hatched and driven by Hyman G. Rickover, a revered yet complicated character who was serving as rear admiral in the U.S. Navy at the time. He was born in 1900 in a small town not far north of Warsaw. His father, who was a tailor, left Europe and brought his family to New York in 1906, when Rickover was six years old. The speed with which the U.S. Navy was able to build a functional submersible vessel powered by a nuclear reactor was the direct result of Rickover's "daring aggressiveness," according to Kintner—a breakthrough that had the potential to transform a submarine into something more than a "surface ship which could submerge only for short periods," but rather into an underwater vessel that would be able to remain hidden in the depths for months.

Rickover could be condescending and abusive. On several occasions, he reportedly made junior officers with whom he disagreed stand in a closet for hours to contemplate their perceived failings. Rickover understood his own limitations to a great degree; he said he had "the charisma of a chipmunk" in an interview with Diane Sawyer on *60 Minutes* in 1984. In his mind, the rules were for other people. When a deputy arrived in his office with a book of U.S. Navy regulations, Rickover recalled telling the officer to get out and burn the book. "My job was not to work within the system. My job was to get things done," he said. Jimmy Carter, who had served under Rickover as a junior officer in the navy in the late 1940s, decades before running for and winning the presidency, acknowledged that Rickover could be difficult, and even that there had been "a few times, when I hated him." But his reverence for the man was steadfast. Carter would add that aside from his own father "no other person has had such a profound impact on my life."

In the early 1980s, a few years after his retirement, it emerged that Rickover had been accepting a range of gifts and favors for nearly two decades from General Dynamics Corporation, one of the country's leading shipbuilders. A report in 1985 by a U.S. Navy review board concluded that he had received, and often requested, a total of $67,628 worth of gifts from the company over a sixteen-year period, or roughly $4,200 per year from 1961 to 1977. The roster of gifts was eclectic and odd. It included a pair of earrings and a jade pendant valued at $1,125, but also twelve fruit knives with handles made of water buffalo horn, the dry cleaning on frequent occasions of Rickover's suits, a used *Encyclopaedia Britannica* set, eleven hot plates and metal pots for cooking custards, twelve shower curtains, teak trays made from the wood deck of the *Nautilus*, 240 coffee mugs over the years, and eighty-eight paperweights from Tiffany & Co. The roster of items represented a sort of collection of corporate detritus, a smattering of essentially holiday gifts and gestures that any one of which in isolation could possibly have been argued to be minor and de minimis but in aggregate suggested to some an overly comfortable relationship with a defense contractor. Rickover admitted to accepting the gifts and said that many were passed on to others in Congress who supported his efforts. The acceptance of such gifts, ranging from trinkets and mementos to jewelry, was relatively commonplace at the time—a relic of an era when shipbuilders and senior defense officials often saw themselves as partners collaborating against their antagonists and adversaries within the military and in Congress. Rickover would later argue that he could have "made a fortune in the private sector," retiring in 1952, but instead stayed on with the navy for three more decades.

The U.S. Navy concluded that the misconduct merited a warning letter, rather than a formal disciplinary proceeding. But Rickover's enemies, of which there were many, saw the revelations as an opportunity to tarnish the reputation of someone they believed had flown too close to the sun. John Lehman, the secretary of the U.S. Navy at the time the "trinkets" scandal broke and a longtime opponent of Rickover's, said in 1985 that the episode represented a "fall from grace" for the retired admiral. An editorial in the *New York Times* that same year argued that the gifts reflected Rickover's "belief that he was above the rules"—a belief that had "helped him to high accomplishments, but fostered deep flaws of judgment." Some saw an aging admiral who should have simply retired decades before he did.

A lonely few came to Rickover's defense. William Proxmire, then a U.S. senator from Wisconsin, summarily brushed away the allegations against his longtime friend, who Proxmire said "will be known as the father of the nuclear Navy and an indomitable fighter against defense contract abuses long after the petty figures who now run the Navy are forgotten." Rickover was, by nearly every account, a towering figure, without whom the United States might never have attained such a decisive advantage over the Soviet Union, tipping the balance of power in America's favor. An obituary in *Time* magazine concluded that while he had been "marred by an excess of arrogance," it was his "rude genius" that "proved to be one of the Navy's greatest assets at the dawn of the Atomic Age."

. . .

The Rickovers of society, and there have been many over the decades and indeed centuries, have for the most part been cast out, discarded as relics of an era when those in power justified, both to themselves and to others, their own self-dealing and mercenary tactics by their ability to achieve results. We have, as a culture, decided to shift our focus to the enforcement of the administrative rules and regulations that many tell themselves are our best and perhaps only defense against a slow decline into corruption. Yet we refuse to engage with what is lost and traded away—the preservation of some degree of space for those whose intentions are noble enough and, more important, whose interests are aligned with those of the group. The speed with which we increasingly have abandoned the unpopular, the unlikable, and the less than charismatic personalities among us should give us pause. The risk is that we begin to privilege the seemingly unobjectionable goals of transparency and process over what actually matters—building submarines, developing our most elusive cures, preventing terrorist attacks, and advancing our interests. Such a utilitarian calculus is unattractive. But in any struggle, we must sometimes set aside aesthetic distaste. We too often hide behind our piety as a way of avoiding more challenging and indeed uncomfortable questions about outcomes and results.

The world looks the other way when confronted with the princely sums paid to those in Silicon Valley and on Wall Street, as well as the hedge fund managers and traders who allocate capital in our market economy. But an uproar arises when a retired navy admiral, one whose efforts provided us with the most significant development in naval warfare of the century, reveals his vanity and lack of judgment when dealing with a defense contractor. Had he broken the rules? Perhaps. But there are costs as well to such a strict and unwavering adherence to such protocols, and limits to the comfort that a narrow procedural justice can provide. Our desire for purity is understandable. We cling to the hope that the most noble and pious among us will also have the ambition to seek power. But history tells us that the opposite is far more often the case.* The eradication of any space for forgiveness—a jettisoning of any tolerance for the complexities and contradictions of the human psyche—may leave us with a cast of characters at the helm we will grow to regret.

The collective desire for a scapegoat can be so thorough and complete that it often, throughout history, has overtaken us. In *Permanence and Change*, published in 1935, Kenneth Burke described "the scapegoat mechanism in its purest form," as "the use of a sacrificial receptacle for the ritual unburdening of one's sins." This process of transferring the sins of a people to an animal, which would then be "ferociously beaten or slain," was a means of relieving the broader social group of guilt or feelings of dissonance. We must grapple far more directly with this cyclical and deeply seated desire that wells up in us for a scapegoat—a vessel for our own failings, weaknesses, forbidden desires, and flaws. The feelings of relief and unburdening that accompany the sacrificial slaughter of the animal, or one of us in our midst, are often ephemeral.

Our society has grown too eager to hasten, and is often gleeful at, the demise of its enemies. The vanquishing of an opponent is a moment to pause, not rejoice. In the sixth century, in a small village outside Rome, Saint Benedict found himself harassed and persecuted by a priest named Florentius. The Roman Empire had collapsed a century before, and Benedict had fled the former imperial capital to pursue a new monastic life in the countryside. Florentius, after attempting to kill Benedict, including by sending him a loaf of poisoned bread that a crow took and cast away, sent "seven naked girls" into the garden of his monastery in a bid to tempt the monk to sin, according to an account of the episode written by Pope Gregory

---

* The hope that a governing class will emerge, as our reluctant leaders drafted into service nearly against their will, is ancient. Plato, in *The Republic*, claimed that "good men will not consent to govern for cash or honours," for "they aren't ambitious."

in the sixth century. The plan failed. Florentius was himself eventually killed; the circumstances of his death remain unclear. But when an apprentice rushes to tell Benedict of his enemy's demise, Pope Gregory recounts that Benedict took the news "very heavily, both because his enemy was dead and because his disciple rejoiced thereat."

* * *

Our current tendency toward the prizing of strict adherence to certain norms and regulations is evidence of a more fundamental challenge that our society faces. A rigidity in our approach to addressing malfeasance, and willingness to overlook results, to persecute the unpopular, are symptoms of dysfunction within a society whose leaders have become untethered from the outcomes with which they are purportedly charged to advance. Many no longer share in either the risk or the reward of their decisions. And yet the reshaping of our most critical institutions, along with the incentives we provide to those who lead them, will not be possible without an even more ambitious, and significant, shift. The reconstruction of a technological republic will, in the end, require the resurrection and re-embrace of a sense of national and collective identity that has, throughout history, provided the bedrock for human progress.

Chapter Seventeen

The Next Thousand Years

IN 1993, ROBIN DUNBAR, a British anthropologist, attempted to calculate the maximum number of individuals with whom a person could plausibly maintain functional social relationships. He surveyed the size of bands of humans who live in hunter-gatherer societies, from southern Africa to New Guinea to northern Canada, and came up with an average of 148.4 individuals per group, with the smallest community studied having 90 members and the largest 221. The figure, more often rounded to an even 150 people, has come to be known as Dunbar's number, and represents a sort of theoretical upper limit to the size of a human community whose members maintain direct contact and relationships with everyone else. The Hutterites, for example, descendants of Protestants from Switzerland and elsewhere in central Europe who sought refuge across the American Midwest and Canada in the nineteenth century, themselves identified 150 as the upper bound of the size of a farming community, and a report from the U.S. Department of the Interior from the early 1980s notes that when a group within a Hutterite enclave reaches 130 to 150 individuals, "a daughter colony splits off from the parent." Similarly, a study from the early 1980s documented a community of 198 individuals living in the remote mountains of East Tennessee, nearly all of whom considered themselves related to some degree. Dunbar, who was born in Liverpool in 1947 and taught at Oxford, has noted that the rough upper bound of 150 individuals seems to operate in other contexts, including the size of military formations within the Roman army as well as modern business units in companies.

The task of maintaining human communities with significantly more than 150 or so individuals, of forming direct social relationships and lasting bonds with that many people, is exceedingly difficult. The monkeys and great apes of the world instinctively groom and comb the hair of other members of their groups as a means of establishing social bonds. The trouble is that grooming dozens let alone hundreds of other individuals on a regular basis requires a very significant investment of time and creative energy. For humans, language, principally, fills the gap, allowing us to form substantial connections, real but more often imagined, with far greater numbers of people. The nations of the world, and our sense of national identity or national culture, have been made possible by both spoken and written language—allowing strangers to build with collective purpose and for the public, not merely private, good. Without those "imagined linkage[s]," in the words of the political scientist Benedict Anderson, the tether between individuals who will almost certainly never meet or know one another directly, nothing of the modern era—from medicine to cities to artificial intelligence—would be possible.

But what sustains communities of individuals that number in the thousands or tens of thousands, millions, and even billions? What is capable of binding us together, of offering some degree of cohesion and common narrative that might allow large groups to organize around something other than our own subsistence? It is, without any doubt, some blend of shared culture, language, history, heroes and villains, stories, and patterns of discourse.

Yet identification of anything approaching a national culture, or values, has in recent decades become increasingly fraught and problematic. In 2017, Emmanuel Macron, the French president, gave an address in which he said, "There is not *a* French culture. . . . There are cultures in France." The remark sparked a round of furious debate in the country, with Macron wading into a discussion that has structured life not only in Europe for nearly half a century but also in America. His denial of the existence of a single French culture, while attempting to highlight the cultural diversity of the newly cosmopolitan country, struck at the heart of French identity. Yves Jégo, the mayor of Montereau-Fault-Yonne, a town on the Seine on the outskirts of Paris, fired back at Macron in an essay in *Le Figaro*, critiquing the president's stance as "contrary to the spirit of our republic." Jégo made clear that the aspiration to preserve something in common did not require a claim of superiority, and it did not deny that all cultures are in a process of constant change. His point was instead that abandoning hope of preserving a national and shared culture risks "losing ourselves in materialism." The irony is that those often most skeptical of the market, and the massive inequities that result from a headlong embrace of capitalism, often fail to appreciate that their own distaste for defending culture or concepts of nationhood leaves a void that the market itself fills.

We, in America and more broadly in the West, have for the past half century resisted defining national cultures in the name of inclusivity. But inclusion into what? We have so hollowed out the national project that one could argue that there is no longer much of substance into which anyone might be included. A call today for affirming an American culture, something greater than its constituent parts, risks being cast as divisive and retrograde. Our sense of civic affiliation with one another has been allowed to wither, and other means of fulfilling that desire for interpersonal tethers have emerged, to fill the yawning gap, including the sense of belonging and investment in a grand narrative of triumph and defeat that can be found, for instance, in sport. Such allegiances will emerge. We will find a way to build coalitions and bands of warriors. To deny the human need for such affiliation has been a mistake.

No country in the history of humanity has done more than the

FIGURE 13
Support for U.S. Major League Baseball Teams as of 2014

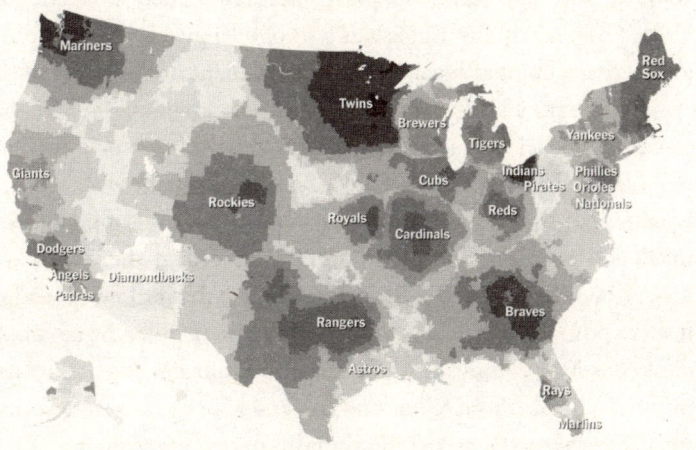

United States, imperfect as it may be, to construct a nation in which membership means something more than a shallow appeal to ethnic or religious identity. Are we to abandon any attempt at building on and expanding that project? The United States, nearly two and a half centuries after its founding, remains defined in part by its contradictions. But other countries, including some of history's most vaunted democracies, continue to struggle with adopting a less parochial conception of national character. In June 1996, Jean-Marie Le Pen, then the president of the National Front party in France, dismissed the country's football team as "a bit artificial," given the number of players who, while French citizens, were descendants of individuals from overseas territories and Africa.

The experience of living in the United States, for many, has grown too fractured, too disparate to allow for such a broad aspiration to something common and shared. It is almost as if Americans have ceded their ability to draft the country's cultural history to others, abandoning space for any such discussion to the editors of foreign textbooks *about* America—to histories being written by others from the outside looking in. Indeed, the editors of the textbook *American Culture*, published in 2008 principally for students outside the United States learning English as a second language, offered a pithy and perhaps unintentionally critical assessment of the status of the American national project: "The study of American culture has moved from being a search for a national character or a national identity to focus on American conflicts, within and without." The issue is that humans will inevitably seek out ways of finding intimacy and connection with strangers, with people they will never meet. Should we challenge the nation's role in that process? Or allow it to step into a breach that would otherwise be filled by an ascendant consumer culture, in which identity and belonging are defined by what one can afford to buy and, as a result, one's caste and wealth? This is, perhaps, the modern left's most glaring strategic mistake. It claims to be committed to curbing the excesses of the market, but its unwillingness to reckon with and take seriously the good that can come from a national culture or shared identity has only enabled the very excesses it purports to oppose.

* * *

On October 3, 1965, Lee Kuan Yew gave a speech at an association of Singapore's liquor retailers, hoping to drum up support for the newly independent nation's cause. It had been only a couple of months since the country split from Malaysia, and Lee was charged with convincing a skeptical public that the island nation had a future on its own. "I am calculating in terms of the next generation, in terms of the next hundred years, in terms of eternity," he said. "And believe you me, for the next thousand years, we will be here." He added, "It is people who calculate and think in those terms that deserve to survive." To many, Singapore's odds of survival after separating from the British Empire and later winning independence from

Malaysia in 1965 were slim. The tiny nation, not much more than an island, lacked the natural resources or population that would seem necessary for any sort of longevity. The country's citizens also spoke nearly a dozen languages and came from distinct cultural and religious traditions, each of which had ancient and deep roots in southern China, on the Indian subcontinent, and across the Malay Peninsula. Lee worked to manufacture some form of national identity for the young country, stitching together what he hoped would become a coherent whole from a diverse array of constituent parts. To that end, he and others unabashedly involved Singapore's government in any number of aspects of the private lives of its citizens, including everything from appropriate manners to the search for a spouse.

At a political rally in 1986, Lee made the case that intervention in the private domains of the country's citizens was a necessary component of constructing and building a nation. "We sang different songs in different languages," he said. "We did not laugh at the same jokes, because you can crack a joke in Hokkien," he added, referring to one of the country's Chinese dialects, but "forty percent of the population won't follow you." For most of the twentieth century, at least twelve Chinese dialects had been spoken in Singapore, including Cantonese, Hokkien, Hainanese, and Shanghainese. The rise and increasing prominence of Chinese dialects in the territory was a relatively recent development. The British colony, through the nineteenth and early twentieth centuries, had emphasized Malay, as opposed to Chinese, given that, as one historian has noted, Singapore was considered "part of a larger Malay world in which Malay was the main lingua franca."

A government review completed in 1979 found that the vast majority of children in the newly independent nation—85 percent—spoke a language other than English or Mandarin at home. The authors of the report wrote, "One of the dangers of secular education in a foreign tongue is the risk of losing the traditional values of one's own people and the acquisition of the more spurious fashions of the west." A shared language was seen as vital to the nation's ability to defend its culture against encroachment and indeed survive over the longer term. "A society unguided by moral values can hardly be expected to remain cohesive under stress," noted the government study, which came to be known as the Goh Report, after its principal author, Goh Keng Swee, Singapore's deputy prime minister under Lee. "It is a commitment to a common set of values that will determine the degree to which people of recent migrant origin will be willing and able to defend their collective interest."

A plan was hatched shortly thereafter to require that all Chinese students learn Mandarin at school instead of the dialects that they had been speaking at home. It was a decisive and controversial move, one with far-reaching consequences for generations of the country's families. "Singapore used to be like a linguistic tropical rain forest— overgrown, and a bit chaotic but very vibrant and thriving," Tan Dan Feng, who served on the country's national translation committee, said in an interview in 2017. "Now, after decades of pruning and cutting, it's a garden focused on cash crops: learn English or Mandarin to get ahead and the rest is useless, so we cut it down."

For his part, Lee continued to make the case that learning Chinese, and an ability to converse with citizens across the country, was essential for the psychological development and coherence of young Singaporeans of Chinese descent. And many credit Lee for essentially rescuing the nation from devolving into a clash of competing bands formed along ethnic or linguistic loyalties. Saravanan Gopinathan, a former dean at the National Institute of Education in Singapore, wrote in 1979 that the country's language policies were instrumental in constructing and maintaining "the cultural personality of the nation." Lee later considered relaxing his grip on the country's development in certain limited domains. "This is a new phase," he explained at the National Day rally in 1986. "Give them the option. You decide. You make up your mind. You exercise the choice. You pay the price." The ascent of Singapore, whatever the mix of causes that propelled its rise, has been undeniable. In 1960, Singapore's per capita gross domestic product was only $428. By 2023, its GDP per capita had risen to $84,734—one of the steepest and most unrelenting climbs of any country in the twentieth century and perhaps in modern history.

. . .

Few, if anyone, could take issue with the view that a single individual, Lee, was absolutely critical to Singapore's rise over its first half century of existence. As Henry Kissinger put it, in the case of Lee's leadership, "the ancient argument whether circumstance or personality shapes events" was "settled in favour of the latter." That ancient argument had stretched back to at least the nineteenth century, when Thomas Carlyle, a Scottish historian, wrote in 1840 of "the Great Man" who had "been the indispensable saviour of his epoch;— the lightning, without which the fuel never would have burnt." The view that lone individuals were the principal drivers of history was common at the time. The Panthéon in Paris, which was built in the eighteenth century to house the remains of the country's most distinguished politicians, philosophers, and generals, includes sculptures of Voltaire, Rousseau, and Napoleon, in a pediment above twenty-two soaring and imposing Corinthian columns. An inscription in the stone, in large capital letters, is legible from the street: "Aux Grands Hommes La Patrie Reconnaissante" (To the Great Men, the Grateful Nation).

A singular emphasis on the acts and thoughts of lone individuals, in assessing a sweep of human affairs that was also driven by economic and political forces, among others, was undoubtedly misplaced. Many may also be unable to look past the reference to men at the exclusion of women. But why are we incapable of disavowing the sexism and parochial sentiment without jettisoning any sense of the heroic as well? Our shift away, as a culture, from this type of thinking, from veneration of leaders, is both a symptom and a cause of our current condition. We have grown weary and skeptical of leadership itself; the heroic has for most gone the way of the mythological—relics of a past that we tell ourselves are irredeemably rooted in a history of domination and conquest. The loss of interest in this way of thinking, narrow and flawed as it was, coincided with the culture's broader abandonment of much interest in character or virtue—seemingly ineffable concepts that could not be reduced to the psychological and moral materialism of the modern age. Our mistake, however, was to throw everything out, instead of simply the bigotry and narrow-mindedness.

The essential failure of the contemporary left has been to deprive itself of the opportunity to talk about national identity—an identity divorced from blood-and-soil conceptions of peoplehood. The political left, in both Europe and the United States, neutered itself decades ago, preventing its advocates from having a forceful and forthright conversation about national identity at all—an identity that might have been linked to a culturally specific set of historical antecedents but rose up beyond them to encompass those who were willing to join. Indeed, a generation of academics and writers refused to patrol the boundaries of the emotional nation at all—the imagined community of Anderson. Richard Sennett, a sociology professor at the London School of Economics, suggested that it may be possible to find "ways of acting together" without relying on what he described as "the evil of a shared national identity." The political philosopher Martha Nussbaum similarly castigated "patriotic pride" as "morally dangerous," urging that our "primary allegiance" should be "to the community of human beings in the entire world." Their project, essentially, was post-national. That move, however, toward an abolition of the nation was ill-advised and premature, and the left has been slow in recognizing its mistake.

• • •

In 1882, Ernest Renan, a French philosopher who was the descendant of fishermen, delivered a speech at the Sorbonne in Paris that was titled "Qu'est-ce qu'une nation?" ("What Is a Nation?"). He was among the first writers to attempt to distinguish the concept of a nation from a more limited or narrow sense of ethnic or racial identity, noting the "graver mistake" occurs when "race is confused with nation." Renan gave voice to a far more enduring and robust concept of the nation, that grand and mysterious collective project, in a way that the educated class all but abandoned in the postwar period. He described the nation as "a vast solidarity, constituted by the sentiment of the sacrifices one has made and of those one is yet prepared to make." A national project, for Renan, "presupposes a past," but is "summarized in the present by a tangible fact: consent, the clearly expressed desire to continue a common life." It is that "common life" with which we are at risk of losing touch. Renan famously described the nation as "an everyday plebiscite." And it must now be renewed.

The necessary task of building the nation, of constructing a collective identity and shared mythology, is at risk of being lost because we grew too fearful of alienating anyone, of depriving anyone of the ability to participate in the common project. It is this disinterest in mythology, in shared narratives, that we have as a culture taken too far. Palantir takes its name from *The Lord of the Rings*, by J. R. R. Tolkien, and some have suggested that Tolkien references are favorites of the "far right." The critique is representative of the left's broader error, both substantive and strategic. An interest in rooting the aims of a corporate enterprise in a broader context and mythology should be celebrated, not dismissed. We need more common tomes, more shared stories, not fewer, even if they must be read critically over time.*

---

* See, for example, an essay by Rowan Williams, the former archbishop of Canterbury: "Master of His Universe: The Warnings in JRR Tolkien's Novels," *New Statesman*, August 8, 2018.

Such stories, the parables and small myths that animate and make possible a larger life, will find refuge in other domains if we continue to insist on excluding them from our civic and public lives. Randy Travis, whose melodies spurred a sort of neoclassical revival in country music in the 1980s and 1990s, recounted tales that had been cast out by American culture as facile and nearly regressive. His song "Three Wooden Crosses," which told the story of "a farmer and a teacher, a hooker and a preacher," epitomized the type of parable that no longer quite fit within ascendant elite culture—an unabashed and unironic account of virtue and redemption. Yet Travis, and his music, remain immensely popular among certain swaths of the public. Our yearning for story and meaning has not withered. It has rather been forced to find expression in domains other than the civic.

• • •

The challenge is that a commitment to participating in the imagined community of the nation, to some degree of forgiveness for the sins and betrayal of one's neighbor, to a belief in the prospect of a greater and richer future together than would be possible alone, requires a faith and some form of membership in a community. Without such belonging, there is nothing for which to fight, nothing to defend, and nothing to work toward. A commitment to capitalism and the rights of the individual, however ardent, will never be sufficient; it is too thin and meager, too narrow, to sustain the human soul and psyche. James K. A. Smith, a philosophy professor at Calvin University, has correctly noted that "Western liberal democracies have lived off the borrowed capital of the church for centuries." If contemporary elite culture continues its assault on organized religion, what will remain to sustain the state? What have we built, or aspired to build, in its place? It is true, as Robert N. Bellah wrote in 1967, that there "exists alongside of and rather clearly differentiated from the churches an elaborate and well-institutionalized civil religion in

America." He made the argument that "this religion—or perhaps better, this religious dimension—has its own seriousness and integrity and requires the same care and understanding that any religion does." A loose constellation of "biblical archetypes," as Bellah put it, including stories from Exodus and sacrifice as well as resurrection, may be a start, but we have grown skeptical and dismissive of even those modest references in public life.

The leaders of Silicon Valley are drawn from a disembodied generation of talent in America that is committed to little more than vehement secularism, but beyond that nothing much of substance. We must, as a culture, make the public square safe again for substantive notions of the good or virtuous life, which, by definition, exclude some ideas in order to put forward others. It is the "pluralism which threatens to submerge us all," as the moral philosopher Alasdair MacIntyre has written, that must be resisted. It is now time, as he made clear, to construct "new forms of community within which the moral life" can "be sustained."

An aspirational desire for tolerance of everything has descended into support of nothing. The contemporary left establishment inhabits a prison of its own making. Like a caged animal, it is left to pace furtively, unable to offer an affirmative vision of a virtuous or moral life, whose content it long ago stripped away to the bare essentials. We must instead now conjure a new "resolve," as the author and art critic Roger Kimball has written, and indeed "self-confidence, faith in the essential nobility of one's regime and one's way of life."

• • •

In 1998, the German Publishers and Booksellers Association decided to award its international peace prize to Martin Walser, one of the country's leading writers and public intellectuals. Walser was born in 1927 in Wasserburg am Bodensee, a town on the shore of Lake Constance, which sits at the southern end of Germany and borders Switzerland and Austria. His parents were Catholic, and he grew up just as Hitler was coming to power in the 1930s. It would later emerge that he joined the Nazi Party when he was seventeen years old, according to reporting by a German magazine that had obtained a 1944 party registration card with Walser's name from the German federal archives in Berlin. Walser told the magazine that he had likely been added to a party roster without his knowledge. He was eventually recruited to the German army and served under Hitler's command through the end of the country's defeat by Allied forces in 1945.

His complexity as a literary and moral figure was perhaps part of his appeal to the German public, and to the publishers' association that had awarded him the peace prize that year. For decades, the country had been subsumed by moral debates and furtive efforts to construct an industry of remembrance of Germany's descent into darkness in the late 1930s and the 1940s. A certain exhaustion had taken hold, and the public, many of whom by that point had been born well after the end of World War II, had grown confused and fatigued by reminders of a horror in which their parents or grandparents, but not themselves, had participated.

At his speech in St. Paul's Church in Frankfurt in October 1998, Walser departed from the standard script of self-flagellation and dutiful acceptance of what many believed was a nation's collective guilt and responsibility. Instead, he suggested that the yoke of an enforced remembrance should be thrown off and abandoned—that the imposition of shame on a contemporary German public had ceased to serve any productive purpose. Walser said, "Everyone knows the burden of our history, our everlasting disgrace." He did not, however, stop there. The daily reminders of Germany's past, for Walser, were more of a self-serving attempt by the country's elite to relieve "their own guilt" than anything else. Walser confided to the audience that he had found himself turning away, refusing to look, at the images of brutality that had become a routine part of German television programming at the time. He explained, "No serious person denies Auschwitz; no person who is still of sound mind quibbles about the horror of Auschwitz; but when this past is held up to me every day in the media, I notice that something in me rebels against this unceasing presentation of our disgrace." Walser denounced efforts to, in his words, trivialize Auschwitz, to make it "a routine threat, a means of intimidation or moral bludgeon." A commentator at the time noted that for Walser the moral failure of a nation had "been instrumentalised by large sections of the media," as well as a "dominant left-liberal intelligentsia as a means of defying German national identity."

The audience during Walser's speech that day included some of the most prominent figures of "the political, economic and cultural German elite," an observer would later write. Roman Herzog, the German president, was in attendance, along with members of the publishing and financial industries. The moment was deeply cathartic for nearly everyone listening, who, according to several accounts, stood up at the end of Walser's speech to give the author sustained applause. He had articulated the forbidden desires and feelings of a nation, and in doing so relieved an immense amount of internal dissonance for his audience, most of whom had been immersed in a culture in which speech had been tightly patrolled and monitored for even the slightest signs of deviation from the received wisdom, the national consensus.

A lone figure in the audience that day declined to stand and applaud. Ignatz Bubis, the chair of the Central Council of Jews in Germany and a towering figure of moral authority in the country, believed that Walser's remarks, while strenuously couched in language aimed at providing cover against charges of antisemitism, were essentially divisive, threatening to take the country back, not forward. The day after the speech, Bubis issued a statement to the

German press accusing Walser of "spiritual arson," or *geistige Brandstiftung*. The two, Walser and Bubis, engaged in a lengthy public debate that captivated the public, with dueling factions lobbying for either holding on to the past or letting it go.

For us, today, the episode provides a reminder of the discomfort and challenges in pressing forward with the task of stitching together something shared from the disparate strands of individual experience. An intense skepticism of German identity, of allowing any sense of the nation to take hold in the wreckage of the war, has had significant costs and deprived the continent of a credible deterrent to Russian aggression. The dismantling of a German national project was, of course, necessary after its descent into madness in the 1930s and 1940s. But many have strained to ensure that nothing quite substantial is permitted to rise from the ashes. This is a mistake, and one that we, in America and other countries, are at risk of repeating. Our persistent unease with broader forms of collective identity must be set aside. To abandon the hope of unity, which itself requires delineation, is to abandon any real chance of survival over the long and certainly very long term. The future belongs to those who, rather than hide behind an often hollow claim of accommodating all views, fight for something singular and new.

Chapter Eighteen

An Aesthetic Point of View

IN 1969, THE TELEVISION SERIES *Civilisation*, a monumental and ambitious account of art history from ancient Rome to medieval France and beyond, aired in the United Kingdom, captivating households across the country. More than two million people watched the program. Church services were rescheduled in some parishes to ensure that the show could be seen. The program's presenter, Kenneth Clark, born in 1903 in London, was the product of another age, unapologetically aristocratic; he offered what appeared to be a coherent narrative of the march toward beauty and greatness of Western art. In postwar Britain, as well as the United States, his worldview was comforting to many and intentionally anachronistic. Clark's judgments about the merits of an artist's work or an epoch's aesthetic bounty had the force of legal edicts. The painting of sixteenth-century Rome struck him as "feeble, mannered," and "self-conscious."

For Clark, there was high and low, and civilization was, or at least should be, on a march toward something greater. He compared an African mask, leaving its country of origin on the continent unspecified, with the Apollo Belvedere at the Vatican, concluding with characteristic assuredness that "the Apollo embodies a higher state of civilisation than the mask." Elsewhere he declined, with a bracing dismissiveness, to provide Spain a central role in the history of

Western civilization, questioning what of significance the country had done "to enlarge the human mind and pull mankind a few steps up the hill." Clark represented, and his work continues to stake out, a certain ideological pole: the view that sweeping aesthetic, and nearly moral, judgments could be made about entire cultures. Their sense of taste, capacity for innovation, and ultimately contribution to human progress were all fair game for assessment and review.

The public consumed his narrative but ultimately revolted. Clark, and his series, have been the subject of sustained attacks in the decades since the program's release. Mary Beard, a British author and historian, recalled in 2016, decades after first encountering the series, that she had begun "to feel decidedly uncomfortable with Clark's patrician self-confidence and the 'great man' approach to art history—one damn genius after the next." So much of what Clark said could never be said today. But in our rush to rebel against the oppression of a narrow account of Western art and history, we perhaps deprived ourselves of more than we anticipated. The sweeping away of anachronisms such as Clark coincided with the abandonment of other normative and aesthetic frameworks. And we have, as a result, unwittingly diminished our capacity to discern and indeed judge.

Even modest attempts to invoke beauty today—such as a swipe at a recent theatrical production by the columnist Peggy Noonan as "ugly, bizarre, inartistic"—are now fraught and resisted. The reshaping of art criticism, the challenges to Clark's mode of being, were the canary in the coal mine. Art might have been first, but much more was to follow. Taste and broader expressions of aesthetic preference—indeed, the suggestion of any preference at all in some contexts—have been shunned as divisive, and mere expressions of elite sensibility. As David Denby wrote in an essay in the *New Yorker* in 1997, "aesthetic taste" is now at risk of being dismissed as a mere product of "status-seeking behavior." It is true, of course, that purportedly neutral or innocent aesthetic decisions are often means of constructing and maintaining caste hierarchies. Thorstein Veblen, an

American sociologist, observed in 1899 that the "circuitous" driveways on the secluded country estates of the British elite, with their gratuitous curves, were a means of expressing power. But is there nothing in our aesthetic lives, no sense of north or south, that ought to be retained?

Our collective and contemporary fear of making claims about truth, beauty, the good life, and indeed justice have led us to the embrace of a thin version of collective identity, one that is incapable of providing meaningful direction to the human experience. All cultures are now equal. Criticism and value judgments are forbidden. Yet this new dogma glosses over the fact that certain cultures and indeed subcultures, including the norms and organizational habits of Silicon Valley, notwithstanding its flaws and contradictions, have produced wonders. Others have proven middling, and worse, regressive and harmful. We are perhaps right to recoil at the summary abandonment of the unnamed "African mask" in favor of the white marble of the Apollo. But should we be left with no means of discerning between art that moves us forward, ideas that advance humanity's cause, and those that do not? The risk is that our fear to pronounce, to speak, to prefer, has left us without direction and confidence when it comes to marshaling our shared resources and talents. Fear has led us to recoil and shrink our sense of the possible, and this fear has found its way into every aspect of our lives.

This abandonment of an aesthetic point of view is lethal to building technology. The construction of software requires taste, both in crafting the programs involved and in selecting the personalities required to build them. It is as much an art as it is a science. Silicon Valley has risen from a small patch of land in Santa Clara County, and built so much and so quickly, in part because it preserved space for the Clarks of the world. Founders have an aesthetic point of view. Their métier might not be nineteenth-century sculpture or Italian frescoes, but they found a space in the Valley that permitted them to exercise what is essentially an artistic form of judgment, and to create in a world where normative claims about good and bad, and narrative arcs of triumph and defeat, were still permitted to exist. The outperformers of the current moment, those who founded and built the world's largest technology companies, which in their size and influence now rival small countries, have largely walled themselves off from society in order to build. Their craft required insulation from the world, not immersion in it, as well as personal judgment and preference.

The commitment to a single path or point of view, and the limiting of one's options, can sometimes be the most effective, indeed the

FIGURE 14
Ulysses and the Sirens by Herbert James Draper, 1909

A number of artists have depicted the Sirens as feminine forms of erotic temptation.*

---

* One art historian has noted that while the Sirens were initially conceived of as "birds with the faces of women" in ancient Greek and Roman artwork of the scene from Homer's *Odyssey*, they eventually became "conflated with mermaids in the Middle Ages."

only, means of navigating the vicissitudes and pressures of public life. When Odysseus asked his crew to tie him to the mast of his ship as it sailed past the Sirens and their bewitching call, he warned his men, "If I should entreat you, and bid you set me free," then "with still more fetters bind me fast." He was intentionally restraining his own range of motion, his ability to respond to the outside world and to the risk of being diverted by its enchanting, and indeed deadly, temptation. A freedom of motion, to maneuver at will, can masquerade as an imitation of power. A willingness to constrain choice, to cast oneself to the mast, is often the best, if not only, route to creative production, for either a company or a culture.

* * *

The outperformance of founder-led companies, for which there is a growing body of evidence, is the result of this privileging of an aesthetic point of view, of space to pronounce and decide. Such economic outperformance has been deeply counterintuitive, even confounding, to many. Companies ruled by committee, with increased oversight and control over management, should, according to the catechism of the free market, be more efficient and effective over time. The evidence, however, suggests otherwise.

Rüdiger Fahlenbrach, a finance professor at École Polytechnique Fédérale de Lausanne in Switzerland, compiled a list of 2,327 U.S. companies over a ten-year period from 1992 to 2002, of which 361 were run by a founder as opposed to a professional or appointed CEO. He found that an investment approach that purchased shares solely in companies run by founders would have earned an excess annual return of 10.7 percent, or 4.4 percent more per year than a portfolio that included all companies, founder-run and otherwise, even when controlling for various other factors including industry and the age of the business. Similar results had been observed with family-owned firms, but Fahlenbrach's research helped distinguish the drivers behind the faster than average growth rates of companies that were controlled by a single family from those that were run by that family as well. He concluded that "a large ownership stake by descendants of a founding family" alone was insufficient to affect a company's value on the market; it was rather firms that maintained a founder at the helm that reliably outperformed over time.

FIGURE 15
The Founder Premium: Total Return of Founder-Led Companies vs. Others (1990 to 2014)

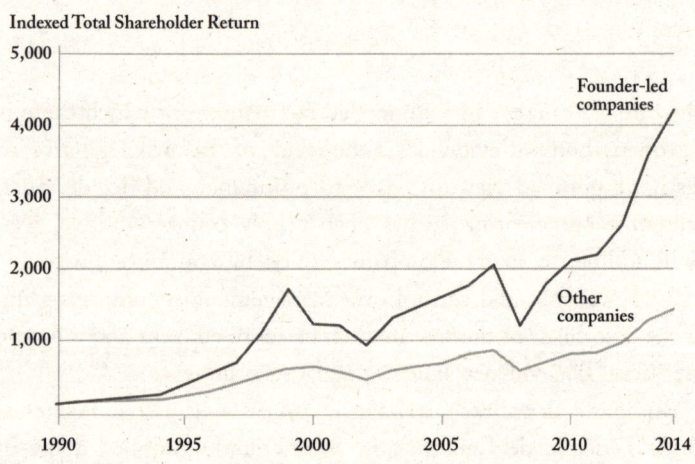

Others have observed similar results. A group of researchers at Purdue University surveyed the five hundred companies in the S&P 500, an index of the largest and most significant businesses in America, over a ten-year period from 1993 to 2003 to determine whether founder-led companies produced more innovation, as measured by patents that were cited widely by others. The researchers were interested not in the mere filing of a patent application but rather in patents that, over time, came to be referenced repeatedly in academic journals and other publications. The team at Purdue found that companies led by founders as opposed to professional CEOs held a 31 percent higher number of significant patents.

Such outperformance is anything but an accident. The union of the pursuit of innovation with the rigor of engineering execution requires a degree of insulation from the outside world, some protection from the instincts and often misdirections of the market. Nothing much of substance, and certainly nothing lasting, will be created by committee. Our challenge, both in the United States and in the West more broadly, will be to harness and channel the creative energies of this new founding generation, these technical iconoclasts, into serving something more than their individual interests.

An ownership culture must be allowed and encouraged to take root in our society. David Swensen, the former investment chief of Yale's endowment, was at the helm of the organization for thirty-five years. He spoke of investing the resources of the university, which was founded in 1701, in order to ensure not years or even decades of strong performance but another three centuries of the school's existence. For Swensen, "short-termism" and the market's "focus on quarter-to-quarter earnings" are "incredibly damaging," as he said in an interview in 2017. It is rather a sort of stewardship, of the temporary and conditional ownership of an asset, that allows one to preserve its value over the long term.

One of the central advantages of Silicon Valley was its embrace—imperfect, halting, and full of contradictions—of an ownership society, a regime in which the labor, the creative talent within organizations, had a substantial stake in the success and outcomes of the businesses they were building. It is easy to forget that the act of granting equity to all employees at a technology company, from administrative assistants to executives, was a radical one in the 1990s, departing from the prevailing model of hourly rates and salaries for an organization's staff while owners reaped outsize rewards. A handful of other industries had flirted with shared ownership models, from law firms to medical practices, but the significant equity stakes were in practice often limited to a thin swath of managers at the helm of an organization.

Silicon Valley went much further, and the strategy proved essential to its success. Many of the world's most prominent technology companies were essentially communally owned. The early participants shared in the risk and the reward. Silicon Valley remains one of the few places in the world where individuals of low birth, to use a phrase of the constitutional law scholar Akhil Reed Amar, can own something substantial and participate in the upside of their labor, rather than remain cogs, even if often highly paid cogs, in the ventures of another. Throughout the 1980s and 1990s, a talented graduate could join Goldman Sachs, which was a pioneer of partnership compensation models, or perhaps a white-shoe law firm, where attorneys shared the profits, and risk, of their work. But those experiments have essentially withered; such firms still attract talented and ambitious minds, but many are paid salaries, often high ones, but salaries nonetheless. The upside of the endeavors and creative energy of labor is captured by the capitalists.

· · ·

In 1934, Ruth Benedict published *Patterns of Culture*, in which she recounted her experience living with and studying preindustrial communities in western Canada, Melanesia, and the Southwest of the United States. She described working toward "a more realistic social faith," one that accounted for the variation across human cultures and cultural practices. But she went further as well, describing "the co-existing and equally valid patterns of life which mankind has created for itself from the raw materials of existence." It was her reference to "equally valid" cultures that would prompt a century of discussion and debate. For several generations of anthropologists, the study of preindustrial societies became a means of elevating them, but also by extension unwittingly exempting them from the realm of moral scrutiny.*

Silicon Valley, in its modern form, was a product of this intellectual tradition, of a cultural and moral agnosticism if not relativism that assiduously avoided anything approaching substantive views about the good life. The effective altruism movement, which took hold in the Valley over the past decade and was advanced by the philosopher Peter Singer, among others, sought to build on the intuitive appeal of ethical universalism—the view that all humans, and indeed some nonhumans, should be considered in our moral calculus. The work of Singer, who was born in Melbourne, Australia, and taught at Princeton for more than two decades, was attractive because it seemed to have solved the puzzle: well-being, whether of humans or sea otters, was all that mattered. But this approach provided cover for a generation to avoid more thorny questions about what constitutes a life well lived, the boundaries and content of national identity, and the human search for meaning. Roger Scruton, a British philosopher, critiqued Singer for adopting "a vacuous utilitarianism" whose elegance and simplicity, alluring as they were, nonetheless reduced experience to a single metric. The founders of many of the leading firms in Silicon Valley were not *immoral*, in this sense; they were simply *amoral*, skeptical of grand belief structures and worldviews and affirmative conceptions of what a collective life could or ought to be.

The entrepreneurs of Silicon Valley do not lack idealism; indeed, they often appear to be brimming with it. But it is thin and can

---

* The conceit of this era of ethnography was that the peoples who were the object of study were fixed in time, essentially lacking the capacity to move or develop through history. Margaret Mead, who was one of Benedict's students, was one of a generation of cultural anthropologists who used what has been described as the "ethnographic present" in recounting the lives of the young women and others about whom she wrote in *Coming of Age in Samoa*, published in 1928. Her subjects not only existed apart from the world but were also "grammatically frozen at the moment she had observed them—*swims, eats, tells, knows*," as the author Charles King has noted.

wither under even the slightest scrutiny. The legions of young founders have for decades now routinely claimed that they aspire to change the world. Yet such claims have grown meaningless from overuse. This cloak of idealism was put on in order to relieve these young founders of the need to develop anything approaching a more substantial worldview. And the nation-state itself, the most effective means of collective organization in pursuit of a shared purpose that the world has ever known, was cast aside as an obstacle to progress.

* * *

Leo Strauss, who was born in Prussia in the late nineteenth century and taught at the University of Chicago, argued that the rejection of the moral point of view was in many ways a precondition of the Enlightenment and indeed the scientific revolution that made Silicon Valley possible, writing that "moral obtuseness," a relinquishing or at least pause in the search for a definition of good and evil, "is the necessary condition for scientific analysis." He was also early to observe that such a clean bifurcation of endeavors, between the scientific and the moral, was in practice far more difficult—that "the value judgments which are forbidden to enter through the front door of political science, sociology or economics, enter these disciplines through the back door." For Strauss, the contemporary social scientist had rejected values in favor of a search for truth and convinced himself that such a distinction was possible. But it was this "indifference to any goal, or of aimlessness and drifting," as Strauss put it, that is the seed of our current nihilism as a culture.

The legions of Silicon Valley entrepreneurs and engineers were the successors to a previous generation of academics who had attempted to hide behind the purportedly neutral pursuit of scientific discovery. The enforced neutrality, first in colleges and universities and later within the technology companies that have constructed the world within which we all now operate, has left us with a hollow republic, something far short of that which we are capable of building. But constructing a technological republic, a rich and thriving and raucously creative communal experiment—not merely the bacchanal of permissive egalitarianism of which Strauss warned—will require an embrace of value, virtue, and culture, the very things that the present generation was taught to abhor.

For Lee Kuan Yew, the aspirational ideal was to become, as Confucius urged more than two thousand years ago, a *junzi*, which has been variously translated as an "exemplary person," or "gentleman." This was someone who is "loyal to his father and mother," "faithful to his wife," "brings up his children well," and is a "loyal citizen of his emperor," Lee said in an interview. Such a specific conception of virtue, for many today, must be resisted as parochial and exclusionary. But what virtues, what conception of the noble or indeed exemplary life, are we willing to advance and defend in the place of the ones that have been jettisoned in the name of inclusivity?

The fall of empires has been accompanied before by an abandonment of the pursuit and nurturing of virtue. Sallust, the Roman historian born in 86 BC, chronicled the descent of the republic around him, as Catiline attempted a coup and was later killed by the Roman army. Sallust wrote that "as a result of riches, the youth were suddenly consumed with luxury and greed, together with insolence." And they grew disinterested in anything beyond their own enrichment. A bland and unsatisfying utilitarianism will not suffice to remedy the current malaise. The effective altruists were shrewd in co-opting the language of moral philosophy, but their move merely delayed a reckoning with man's search for meaning. As Irving Kristol, who founded the *National Interest* in 1985, has written, "The delicate task that faces our civilization today is not to reform the secular, rationalist orthodoxy," but rather "to breathe new life into the older, now largely comatose, religious orthodoxies."

And it is here that the establishment left has failed its cause and so thoroughly eroded its potential. The frenetic pursuit of a shallow egalitarianism in the end hollowed out its broader and more compelling political project. The right was pursued while the good was abandoned. What we need is more cultural specificity—in education, technology, and politics—not less. The vacant neutrality of the current moment risks allowing our instinct for discernment to atrophy.* We must now take seriously the possibility that it will be the resurrection of a shared culture, not its abandonment, that will make possible our continued survival and cohesion.

It was a distaste for collective experience and endeavor that made America, and American culture, vulnerable to attack and infiltration. We had been trained to be so careful, so reluctant to speak about the content of American culture, if there was any, that the act of cultural production and manufacture migrated to other, less hostile domains. At present, the principal features of American society that are shared are not civic or political, but rather cohere around entertainment, sports, celebrity, and fashion. This is not the result of some unbridgeable political division. The interpersonal tether that makes possible a form of imagined intimacy among strangers within groups of a significant size was severed and banished from the public sphere. The old means of manufacturing a nation, the civic rituals of an educational system, mandatory service in national defense, religion, a shared language, and a thriving and free press have all but been dismantled or withered from neglect and abuse.

Silicon Valley seized an opportunity created by the void that had opened in American national experience. The technology companies that have come to dominate our lives were in many cases small na-

---

* John Rawls contended that an aspiration for political liberalism to remain "neutral in aim" did not preclude the possibility that it could "still affirm the superiority of certain forms of moral character and encourage certain moral virtues." But his roster of virtues, including "fair social cooperation," "civility," "tolerance," and "reasonableness," has proven limited and modest—a collection of essentially unobjectionable background requirements for the operation of civil society that did not allow for any richness or cultural specificity in public life.

tions, built around a set of ideals that many young people craved: freedom to build, ownership of their success, and a commitment above all to results. The Sunnyvales, Palo Altos, and Mountain Views of the world were company towns and city-states, walled off from society and offering something that the national project could no longer provide.

Our argument is that the path forward will involve a reconciliation of a commitment to the free market, and its atomization and isolation of individual wants and needs, with the insatiable human desire for some form of collective experience and endeavor. Silicon Valley offered the latter and the rewards of the former. Across the towns, corporate and otherwise, in Santa Clara County, a form of modern-day artistic colony, or technological commune, sprang up. These were internally coherent communities whose corporate campuses attempted to provide for all of the wants and needs of daily life. They were at their core collectivist endeavors, populated by intensely individualistic and freethinking minds. It is true that the communal experience that Silicon Valley firms were selling itself became commoditized. Yet the atomization of daily life in America and the broader West left a lane open for technology firms, including ours, to recruit and retain a generation of talent that wanted to do something other than tinker with financial markets or consult.

Other nations, including many of our geopolitical adversaries, understand the power of affirming shared cultural traditions, mythologies, and values in organizing the efforts of a people. They are far less shy than we are about acknowledging the human need for communal experience. The cultivation of an overly muscular and unthoughtful nationalism has risks. But the rejection of any form of life in common does as well. The reconstruction of a technological republic, in the United States and elsewhere, will require a re-embrace of collective experience, of shared purpose and identity, of civic rituals that are capable of binding us together. The technologies we are building, including the novel forms of AI that may challenge our present monopoly on creative control in this world, are themselves the product of a culture whose maintenance and development we now, more than ever, cannot afford to abandon. It might have been just and necessary to dismantle the old order. We should now build something together in its place.

Acknowledgments

The writing of this book, for both of us, has been a privilege. We are indebted to many collaborators and conspirators over the years, intellectual influences and antagonists that have made possible and shaped our thinking, including in Frankfurt and New Haven, as well as Palo Alto and New York.

The project began with encouragement from Alexandra Wolfe Schiff and a decisive introduction to Sloan Harris. He took a risk in venturing toward, not away, from the book, which resisted categorization, falling in the interstitial but we hope to think rich space between political, business, and academic treatise. His guidance, particularly for two first-time authors, was critical.

A book can, of course, take many forms, depending on the instincts and inclinations of an editor. We were fortunate to have found ours, Paul Whitlatch, who consistently and without hesitation pushed us toward the production of something substantial and ambitious. His sense for the sound of prose, as well as the ways in which we might provoke an authentic discussion, were essential to crafting our argument. We are more than grateful as well for the support of Gillian Blake and David Drake at Crown—both of whom were unwavering in their desire to produce a work that aspired to prompt genuine engagement with the reader, particularly in a publishing industry where those in business are rarely permitted, let alone encouraged, to venture outside of their expected genre.

We are indebted to the research assistance and thoughtful guidance of Landon Alecxih, Bill Rivers, Jack Crovitz, and Sam

Feldman, as well as the counsel and support of Nikolaj Gammeltoft and Julia O'Connell—all of whom were vital in their own ways to ensuring that the manuscript matured, and briskly, into a final product.

• • •

Above all, this book owes its existence to Palantir—the company, its founders, our colleagues, partners, organizational culture, and software. The radical suggestion to build technology that served the needs of U.S. defense and intelligence agencies, instead of merely catering to the consumer, began with Peter Thiel, who sensed the diminished ambition of Silicon Valley and without whom Palantir would not exist. Alex is enormously grateful for his friendship, over more than three decades, and support. The company is also a product and reflection of the creative insight and unwavering commitment of Stephen Cohen, along with the work and dedication of Joe Lonsdale and Nathan Gettings. The construction of something from nothing would never have been possible without the fierce loyalty and leadership of Shyam Sankar, as well as Aki Jain, Ted Mabrey, Ryan Taylor, and Seth Robinson—each of whom has been essential to building what Palantir has become. The unflinching support of others, particularly in the early years and when it was less than fashionable to invest in technology for the defense industry, was vital as well. The partnership of Stanley Druckenmiller, Ken Langone, Marie-Josée and Henry Kravis, and Herb Allen III will never be forgotten.

The ideas expressed here are both an outgrowth of and an attempt at articulating our experience at what we believe is an enormously differentiated institution. It has been a wild and rich experiment indeed, and one for which we are both more than grateful to be a part.

Notes

v **"You will never touch":** See Johann Wolfgang von Goethe, *Faust*, trans. Abraham Hayward and A. Bucheim (London: George Bell and Sons, 1892), 40–41 (for a translation on which this one is based).

vii **"The power to hurt":** Thomas C. Schelling, *Arms and Influence* (New Haven, Conn.: Yale University Press, 2020), 2.

vii **"Fundamentalists rush in":** Michael Sandel, *Liberalism and the Limits of Justice*, 2nd ed. (Cambridge, U.K.: Cambridge University Press, 1998), 217.

This book incorporates and builds on three essays published in the *New York Times* (Alexander C. Karp, "Our Oppenheimer Moment: The Creation of A.I. Weapons," July 25, 2023), *Time* (Karp and Nicholas W. Zamiska, "Silicon Valley Has a Harvard Problem," February 12, 2024), and the *Washington Post* (Karp and Zamiska, "New Weapons Will Eclipse Atomic Bombs," June 25, 2024).

CHAPTER ONE:
LOST VALLEY

3 **This early dependence of Silicon Valley:** See Stuart W. Leslie, "The Biggest 'Angel' of Them All: The Military and the Making of Silicon Valley," in *Understanding Silicon Valley: The Anatomy of an Entrepreneurial Region*, ed. Martin Kenney (Stanford, Calif.: Stanford University Press, 2000), 49 (noting that "missing in virtually every account of freewheeling entrepreneurs and visionary venture capitalists is the military's role, intentional and otherwise, in creating and sustaining Silicon Valley").

3 **In the 1940s, the federal government:** See, for example, Roswell Quinn, "Rethinking Antibiotic Research and Development: World War II and the Penicillin Collaborative," *American Journal of Public Health* 103, no. 3 (March 2013): 427–28 (discussing the development of penicillin);

Lynn W. Kitchen, David W. Vaughn, and Donald R. Skillman, "Role of U.S. Military Research Programs in the Development of U.S. Food and Drug Administration–Approved Antimalarial Drugs," *Clinical Infectious Diseases* 43, no. 1 (July 2006).

3 **Indeed, Silicon Valley once stood:** See also Arthur Herman, *Freedom's Forge: How American Business Produced Victory in World War II* (New York: Random House, 2013) (for an account of the marshaling of American industrial production in service of the country's military aims during World War II); Walter Isaacson, *The Innovators: How a Group of Hackers, Geniuses, and Geeks Created the Digital Revolution* (New York: Simon & Schuster, 2015), 181 (noting that "the first major market for microchips was the military").

4 **Fairchild Camera and Instrument:** Robert Perry, *A History of Satellite Reconnaissance*, vol. 1 (U.S. National Reconnaissance Office, 1973), 43; see also Anna Slomovic, *Anteing Up: The Government's Role in the Microelectronics Industry* (Santa Monica, Calif.: RAND Corporation, 1988), 27.

4 **For a time after:** Thomas Heinrich, "Cold War Armory: Military Contracting in Silicon Valley," *Enterprise & Society* 3, no. 2 (June 2002): 247 (noting that "Santa Clara County ... produced all of the United States Navy's intercontinental ballistic missiles, the bulk of its reconnaissance satellites and tracking systems, and a wide range of microelectronics that became integral components of high-tech weapons and weapons systems").

4 **Companies such as Lockheed:** Heinrich, "Cold War Armory," 248.

4 **In November 1944:** See Traudl Junge, *Until the Final Hour: Hitler's Last Secretary*, ed. Melissa Müller and trans. Anthea Bell (New York: Arcade Publishing, 2004), 145–46.

4 **Both his father:** G. Pascal Zachary, *Endless Frontier: Vannevar Bush, Engineer of the American Century* (New York: Free Press, 1997), 12–13.

4 **In the letter:** Roosevelt to Bush, Washington, D.C., November 17, 1944, in Vannevar Bush, *Science: The Endless Frontier* (Washington, D.C.: U.S. Government Printing Office, 1945), vii.

5 **The challenge was to ensure:** Vannevar Bush, "As We May Think," *Atlantic Monthly*, July 1945, 101.

5 **Many of the earliest leaders:** See, for example, Jon Meacham, *Thomas Jefferson: The Art of Power* (New York: Random House, 2013), 314–15.

5 **Dudley Herschbach:** Walter Isaacson, *Benjamin Franklin: An American Life* (New York: Simon & Schuster, 2003), 129.

5 **For Jefferson, science and natural history:** Jefferson to Harry Innes, Phila-

delphia, March 7, 1791, in *The Papers of Thomas Jefferson*, vol. 19 (Princeton, N.J.: Princeton University Press, 1974), 521.

5 **James Madison dissected:** Madison to Jefferson, June 19, 1786, *The Writings of James Madison*, vol. 2, ed. Gaillard Hunt (New York: G. P. Putnam's Sons, 1901), 249–51; see Lee Alan Dugatkin, "Buffon, Jefferson, and the Theory of New World Degeneracy," *Evolution: Education and Outreach* 12 (2019).

6 **We have, in the modern era:** Smriti Mallapaty, Jeff Tollefson, and Carissa Wong, "Do Scientists Make Good Presidents?," *Nature*, June 6, 2024; Robin Harris, *Not for Turning: The Life of Margaret Thatcher* (New York: Thomas Dunne Books, 2013), 24–25 (noting that Thatcher, as a young student, "wanted to pursue a career and chemistry offered the prospect of a future job in industry" before her turn toward law and politics); Eagleton Institute of Politics, *Scientists in State Politics* (New Brunswick, N.J.: Rutgers University, 2023).

6 **John Adams, the second president:** Edward Handler, "'Nature Itself Is All Arcanum': The Scientific Outlook of John Adams," *Proceedings of the American Philosophical Society* 120, no. 3 (June 1976): 223 (internal quotation marks omitted).

6 **The term "scientist" itself:** Claire Brock, "The Public Worth of Mary Somerville," *British Journal for the History of Science* 39, no. 2 (June 2006), 272; Oxford English Dictionary, 2nd ed. (1989), s.v. "scientist" (citing [William Whewell], review of *On the Connexion of the Physical Sciences*, by Mary Sommerville, *Quarterly Review*, vol. 51 (London: John Murray, 1834), 59).

6 **As of 1481:** François Rigolot, "Curiosity, Contingency, and Cultural Diversity: Montaigne's Readings at the Vatican Library," *Renaissance Quarterly* 64, no. 3 (Fall 2011), 848.

7 **Joseph Licklider:** Robert M. Fano, "Joseph Carl Robnett Licklider," in *Biographical Memoirs*, vol. 3 (Washington, D.C.: National Academies Press, 1998), 200.

7 **His research for his:** J. C. R. Licklider, "Man-Computer Symbiosis," *IRE Transactions on Human Factors in Electronics*, no. 1 (March 1960): 4.

7 **Shortly after the launch:** John S. Rigden, *Rabi: Scientist and Citizen* (New York: Basic Books, 1987), 249; see also Thomas Soapes, "Interview with Hans A. Bethe," Dwight D. Eisenhower Library, Abilene, Kans., November 3, 1977.

7 **In 1942, as war spread:** Zachary, *Endless Frontier*, 149; see "The Press: In a

Corner, on the 13th Floor," *Time*, July 22, 1946 (noting the circulation of *Collier's* that year).

8 **Marie Curie sent a letter:** Eve Curie, *Madame Curie*, trans. Vincent Sheean (Garden City, N.Y.: Doubleday, Doran, 1937), 211.

8 **Similarly, Albert Einstein:** See, for example, *Time*, July 1, 1946 (featuring Einstein on the cover of the issue).

8 **This was the American century:** See Jeremy Mouat and Ian Phimister, "The Engineering of Herbert Hoover," *Pacific Historical Review* 77, no. 4 (November 2008): 582 (noting that in the 1920s and 1930s the entire culture "embraced engineers as heroes," and "engineering achievements—whether bridges, skyscrapers, or common household appliances—became the backdrop of everyday life at the same time that engineers were adopted as symbols of progress and prosperity").

8 **The pursuit of public interest:** See Kenneth Clark, *Civilisation* (New York: Harper & Row, 1969), xvii.

8 **As Jürgen Habermas has suggested:** See Jürgen Habermas, *Legitimation Crisis*, trans. Thomas McCarthy (Cambridge, U.K.: Polity Press, 1976), 46 (describing a "legitimation crisis," where "the legitimizing system does not succeed in maintaining the requisite level of mass loyalty while the steering imperatives taken over from the economic system are carried through").

8 **When emerging technologies:** Habermas explained, "Even if the state apparatus were to succeed in raising the productivity of labor and in distributing gains in productivity in such a way that an economic growth free of crises . . . were guaranteed, growth would still be achieved in accord with priorities that take shape as a function, not of generalizable interests of the population, but of private goals." Habermas, *Legitimation Crisis*, 73.

8 **In this way, the willingness:** See also Raymond Plant, "Jürgen Habermas and the Idea of Legitimation Crisis," *European Journal of Political Research* 10 (1982): 343 (noting that "capitalism has built up expectations about consumption, and these have increased pressures on governments to steer the economy to produce more goods").

9 **The modern incarnation of Silicon Valley:** See, for example, Hamza Shaban, "Google Parent Alphabet Reports Soaring Ad Revenue, Despite YouTube Backlash," *Washington Post*, February 1, 2018 (noting that "revenue from Google's ad business . . . accounts for 84 percent of Alphabet's total revenue").

13 **The void was left open:** See Michael Sandel, *Liberalism and the Limits of Justice*, 2nd ed. (Cambridge, U.K.: Cambridge University Press, 1998), 217.

CHAPTER TWO:
SPARKS OF INTELLIGENCE

16 **For Oppenheimer, the atomic weapon:** Lincoln Barnett, "J. Robert Oppenheimer," *Life*, October 10, 1949, 121.

16 **In high school, Oppenheimer:** Jeremy Bernstein, "Oppenheimer's Beginnings," *New England Review* 25, no. 1/2 (2004): 42 (citing an interview of Oppenheimer by Thomas Kuhn); Kai Bird and Martin J. Sherwin, *American Prometheus: The Triumph and Tragedy of J. Robert Oppenheimer* (New York: Alfred A. Knopf, 2005), 23.

17 **"When you see something":** Richard Polenberg, ed., *In the Matter of J. Robert Oppenheimer: The Security Clearance Hearing* (Ithaca, N.Y.: Cornell University Press, 2002), 46.

17 **At a lecture:** J. Robert Oppenheimer, "Physics in the Contemporary World," *Bulletin of the Atomic Scientists* 4, no. 3 (1948): 66.

17 **Percy Williams Bridgman:** Barnett, "J. Robert Oppenheimer," 133.

18 **Some have predicted:** Ian Bremmer and Mustafa Suleyman, "The AI Power Paradox: Can States Learn to Govern Artificial Intelligence—Before It's Too Late?," *Foreign Affairs*, August 16, 2023 (noting that "'brain scale' models with more than 100 trillion parameters—roughly the number of synapses in the human brain—will be viable within five years"); see also Mustafa Suleyman, *The Coming Wave: Technology, Power, and the Twenty-First Century's Greatest Dilemma*, with Michael Bhaskar (New York: Crown, 2023), 66.

19 **And the most advanced versions:** Sébastien Bubeck et al., "Sparks of Artificial General Intelligence: Early Experiments with GPT-4," arXiv, March 22, 2023, 92.

19 **The computer explained that:** Bubeck et al., "Sparks of Artificial General Intelligence," 11.

20 **The language models are not:** Piotr W. Mirowski et al., "A Robot Walks into a Bar: Can Language Models Serve as Creativity Support Tools for Comedy?," arXiv, June 3, 2024, 2.

20 **In the early 1960s:** Elizabeth Corcoran, "Squaring Off in a Game of Checkers," *Washington Post*, August 14, 1994; see also John Pfeiffer, *The Thinking Machine* (Philadelphia: J. B. Lippincott, 1962), 167–74.

20 **In February 1996:** Kat Eschner, "Computers Are Great at Chess, but That Doesn't Mean the Game Is 'Solved,'" *Smithsonian Magazine*, February 10, 2017.

21 **And in 2015, Fan Hui:** Annie Sneed, "Computer Beats Go Champion for First Time," *Scientific American*, January 27, 2016.

22 **Lemoine was raised on a farm:** Blake Lemoine, "Explaining Google," Medium, May 30, 2019.

22 **Over the course:** Nitasha Tiku, "The Google Engineer Who Thinks the Company's AI Has Come to Life," *Washington Post*, June 11, 2022.

22 **Google fired Lemoine:** Nitasha Tiku, "Google Fired Engineer Who Said Its AI Was Sentient," *Washington Post*, July 22, 2022.

22 **Less than a year later:** Kevin Roose, "Bing's A.I. Chat: 'I Want to Be Alive,'" *New York Times*, February 16, 2023.

23 **Others believed that any shadow:** See, for example, Brian Christian, "How a Google Employee Fell for the Eliza Effect," *Atlantic*, June 21, 2022 (arguing that while the capabilities of the latest large language models are "breathtaking and sublime," the exchanges that "may sound like introspection" are "just the system improvising in an introspective verbal style").

23 **The exchange with Bing:** Peggy Noonan, "A Six-Month AI Pause? No, Longer Is Needed," *Wall Street Journal*, March 30, 2023.

23 **The two transcripts:** See Noonan, "A Six-Month AI Pause?" (adding that an effort by the *Times* columnist "to discern a Jungian 'shadow self' within Microsoft's Bing chatbot left him unable to sleep").

23 **The model, for the skeptics:** Emily M. Bender et al., "On the Dangers of Stochastic Parrots: Can Language Models Be Too Big?," *Proceedings of the 2021 ACM Conference on Fairness, Accountability, and Transparency* (2021): 617.

23 **A professor in the department:** Oliver Whang, "How to Tell if Your A.I. Is Conscious," *New York Times*, September 18, 2023.

24 **Douglas Hofstadter, the author:** Douglas Hofstadter, "Gödel, Escher, Bach, and AI," *Atlantic*, July 8, 2023.

24 **Hofstadter had previously expressed doubt:** James Somers, "The Man Who Would Teach Machines to Think," *Atlantic*, November 2013.

24 **Noam Chomsky has similarly dismissed:** Noam Chomsky, Ian Roberts, and Jeffrey Watumull, "The False Promise of ChatGPT," *New York Times*, March 8, 2023.

24 **The claim made by Chomsky:** See Chomsky et al., "False Promise" (arguing that the "deepest flaw" of the current language models "is the absence of the most critical capacity of any intelligence: to say not only what is the case, what was the case and what will be the case—that's description and prediction—but also what is not the case and what could and could not be the case").

25 **An open letter published:** Future of Life Institute, "Pause Giant AI Experiments: An Open Letter," March 22, 2023.

25 **Eliezer Yudkowsky, an outspoken:** Eliezer Yudkowsky, "Pausing AI Development Isn't Enough. We Need to Shut It All Down," *Time*, March 29, 2023.

25 **Peggy Noonan, in a column:** Noonan, "Six-Month AI Pause?"

25 **Those involved in the debate:** See Sean Thomas, "Are We Ready for P(doom)?," *Spectator*, March 4, 2024; Clint Rainey, "P(doom) Is AI's Latest Apocalypse Metric. Here's How to Calculate Your Score," *Fast Company*, December 7, 2023.

25 **Lina Khan, the head:** Thomas, "Are We Ready for P(doom)?"

25 **Similar predictions, all of which:** See J. McCarthy et al., "A Proposal for the Dartmouth Summer Research Project on Artificial Intelligence," August 31, 1955 (reproduced in *AI Magazine* 27, no. 4 (Winter 2006): 12–14).

25 **At a banquet in Pittsburgh:** Herbert A. Simon and Allen Newell, "Heuristic Problem Solving: The Next Advance in Operations Research," *Operations Research* 6, no. 1 (January–February 1958): 7; see also Garry Kasparov, *Deep Thinking: Where Machine Intelligence Ends and Human Creativity Begins* (New York: PublicAffairs, 2017), 36.

25 **In 1960, only four years:** Herbert A. Simon, *The New Science of Management Decision* (New York: Harper & Brothers, 1960), 38.

26 **He envisioned that:** See Simon, *New Science of Management Decision*, 38 (arguing that "men will retain their greatest comparative advantage in jobs that require flexible manipulation of those parts of the environment that are relatively rough—some forms of manual work, control of some kinds of machinery").

26 **Similarly, in 1964:** Irving John Good, "Speculations Concerning the First Ultraintelligence Machine," in *Advances in Computers*, ed. Franz L. Alt and Morris Rubinoff, vol. 6 (New York: Academic Press, 1965), 78; see also Luke Muehlhauser, "What Should We Learn from Past AI Forecasts?," Open Philanthropy, May 2016 (for a review of past predictions over the latter half of the twentieth century anticipating, incorrectly, the imminent emergence of AI on par with the human mind).

CHAPTER THREE:
THE WINNER'S FALLACY

29 **Rabha makes clear:** The Babylonian Talmud, trans. Michael L. Rodkinson, vols. 7 and 8 (Boston: Talmud Society, 1918), 214; see also Ronen Bergman, *Rise and Kill First: The Secret History of Israel's Targeted Assassinations* (New York: Random House, 2018), 315.

29　**They will proceed:** See, for example, Ross Andersen, "The Panopticon Is Already Here," *Atlantic*, September 2020; Paul Mozur, "Inside China's Dystopian Dreams: A.I., Shame, and Lots of Cameras," *New York Times*, July 8, 2018.

30　**As of 2024:** U.S. National Institute of Standards and Technology, "Face Recognition Technology Evaluation (FRTE) 1:1 Verification," 2024; see also Andersen, "The Panopticon"; Tom Simonite, "Behind the Rise of China's Facial-Recognition Giants," *Wired*, September 3, 2019.

30　**In December 2021:** U.S. Department of the Treasury, "Treasury Identifies Eight Chinese Tech Firms as Part of the Chinese Military-Industrial Complex," December 16, 2021.

30　**Two of the other companies:** U.S. National Institute of Standards and Technology, "Face Recognition Technology Evaluation."

30　**In 2022, a research group:** Xin Zhou et al., "Swarm of Micro Flying Robots in the Wild," *Science Robotics* 7, no. 66 (2022), 3.

30　**A graduate student at École Polytechnique:** Debbie White, "Drones Branch Out to Swarming Through Forests, *Times* (London), May 5, 2022.

30　**Yet the following year:** Emilie B. Stewart, "Survey of PRC Drone Swarm Inventions" (Montgomery, Ala.: China Aerospace Studies Institute, U.S. Air Force, 2023), 20.

31　**He declared months before the fall:** Francis Fukuyama, "The End of History?," *National Interest*, no. 16 (Summer 1989): 4.

31　**Fukuyama's claim:** Allan Bloom, "Responses to Fukuyama," *National Interest*, no. 16 (Summer 1989): 19.

32　**Our point is that such:** Joseph S. Nye Jr., *Soft Power: The Means to Success in World Politics* (New York: PublicAffairs, 2004), 8.

32　**"To be coercive":** Thomas Schelling, *Arms and Influence* (New Haven, Conn.: Yale University Press, 2020), 2.

32　**As he made clear:** Schelling, *Arms and Influence*, 142.

33　**The company had been selected:** Joshua Brustein, "Microsoft Wins $480 Million Army Battlefield Contract," *Bloomberg*, November 28, 2018.

33　**"We did not sign up":** Julia Carrie Wong, "'We Won't Be War Profiteers': Microsoft Workers Protest $48M Army Contract," *Guardian*, February 22, 2019.

33　**"Building this technology":** Scott Shane and Daisuke Wakabayashi, "'The Business of War': Google Employees Protest Work for the Pentagon," *New York Times*, April 4, 2018; see also Jamie Condliffe, "Amazon Is Latest

Tech Giant to Face Staff Backlash over Government Work," *New York Times*, June 22, 2018.

33 **Diane Greene, who ran Google's:** Kate Conger, "Google Plans Not to Renew Its Contract for Project Maven, a Controversial Pentagon Drone AI Imaging Program," *Gizmodo*, June 1, 2018.

33 **An article in *Jacobin*:** Ben Tarnoff, "Tech Workers Versus the Pentagon," *Jacobin*, June 6, 2018.

34 **It is striking:** See "Confidence in Institutions," Gallup (noting that in 2023, 60 percent of respondents to a national survey reported that they had a "great deal" or "quite a lot" of confidence in the U.S. military," as opposed to only 8 percent of respondents who said the same about Congress).

34 **When William F. Buckley Jr.:** Dan Wakefield, "William F. Buckley Jr.: Portrait of a Complainer," *Esquire*, January 1, 1961, 50.

34 **They charge themselves with:** See, for example, Mariana Mazzucato, *The Entrepreneurial State: Debunking Public vs. Private Sector Myths* (London: Anthem Press, 2013), 63 (noting the argument "that while Silicon Valley has been an attractive and influential model for regional development, it has been also difficult to copy it, because almost every advocate of the Silicon Valley model tells a story of 'freewheeling entrepreneurs and visionary venture capitalists' and yet misses the crucial factor: the military's role in creating and sustaining it") (quoting Stuart W. Leslie, "The Biggest 'Angel' of Them All: The Military and the Making of Silicon Valley," in *Understanding Silicon Valley: The Anatomy of an Entrepreneurial Region*, ed. Martin Kenney (Stanford, Calif.: Stanford University Press, 2000)).

35 **There was a degree:** Jean Edward Smith, *FDR* (New York: Random House, 2008), 25; *The U.S. and the Holocaust*, directed by Ken Burns, Lynn Novick, and Sarah Botstein (PBS, 2022).

35 **Einstein and his wife:** Smith, *FDR*, 579; see also "The Roosevelt Week," *Time*, February 5, 1934.

35 **The rapid technical advances:** Einstein to Roosevelt, Peconic, N.Y., August 2, 1939, Franklin D. Roosevelt Presidential Library and Museum, Hyde Park, N.Y.

35 **That permanent contact:** Germany ended up never seriously pursuing an atomic weapon. The country's armaments minister, Albert Speer, concluded in 1942 that "the construction of an atomic bomb was too uncertain and too expensive." Hitler, for his part, was reportedly "uninterested in an atomic weapon and disparaged nuclear science as 'Jewish physics.'" Smith, *FDR*, 580.

CHAPTER FOUR:
END OF THE ATOMIC AGE

37 **The rain, however:** *The Manhattan Project: Making the Atomic Bomb* (Oak Ridge, Tenn.: U.S. Department of Energy, January 1999), 48.

37 **The explosion was described:** Janet Farrell Brodie, *The First Atomic Bomb: The Trinity Site in New Mexico* (Lincoln, Neb.: University of Nebraska Press, 2023), 5.

37 **A government report:** U.S. Department of Energy, *Manhattan Project*, 49.

37 **Nobel, who was born in Stockholm:** Edith Patterson Meyer, *Dynamite and Peace: The Story of Alfred Nobel* (Boston: Little, Brown, 1958), 89.

37 **The industrial chemical:** Meyer, *Dynamite and Peace*, 111.

37 **In the early 1870s:** Meyer, *Dynamite and Peace*, 112.

38 **At first, Nobel intended:** Meyer, *Dynamite and Peace*, 110–12 ("What Alfred Nobel thought of this first wartime use of dynamite, in the Franco-Prussian War, he never, apparently, said—certainly not in writing.").

38 **"The only thing":** Meyer, *Dynamite and Peace*, 196.

38 **John Hersey, the American journalist:** John Hersey, *Hiroshima* (New York: Vintage Books, 2020), 35.

38 **The destruction was total:** Hersey, *Hiroshima*, 40.

38 **Their purpose was to level:** Richard B. Frank, *Tower of Skulls: A History of the Asia-Pacific War: July 1937–May 1942* (New York: W. W. Norton, 2020), 8.

39 **"We hated what we were doing":** John Ismay, "'We Hated What We Were Doing': Veterans Recall Firebombing Japan," *New York Times*, March 9, 2020.

39 **In a dispatch from Berlin:** H. L. Mencken, "Ludendorff," *Atlantic*, June 1917, 825, 830.

39 **For Ludendorff, under the logic:** Erich Ludendorff, *The "Total" War*, ed. Herbert Lawrence (London: Friends of Europe, 1936), 8.

39 **The record of humanity's:** See Eric Schlosser, *Command and Control: Nuclear Weapons, the Damascus Accident, and the Illusion of Safety* (New York: Penguin Press, 2013), 480.

40 **Nearly forty years ago:** John Lewis Gaddis, *The Long Peace: Inquiries into the History of the Cold War* (New York: Oxford University Press, 1987), 245.

40 **Steven Pinker, in his book:** Steven Pinker, *The Better Angels of Our Nature: Why Violence Has Declined* (New York: Penguin Books, 2011), 692.

41 **Any number of other developments:** See Robert Rauchhaus, "Evaluating the Nuclear Peace Hypothesis," *Journal of Conflict Resolution* 53, no. 2 (April 2009), 264.

42 **"Free riders aggravate me":** Jeffrey Goldberg, "The Obama Doctrine," *Atlantic*, April 2016.

42 **"After the Cold War":** Paul Taylor, "How to Spend Europe's Defense Bonanza Intelligently," *Politico*, September 2, 2022.

42 **The implications of the fractured:** "Europe Faces a Painful Adjustment to Higher Defence Spending," *Economist*, February 22, 2024.

43 **In February 1951:** See "Bermingham, Once Chicago Banker, Dies," *Chicago Tribune*, July 14, 1958, 48.

43 **The challenge, as Eisenhower put it:** Eisenhower to Bermingham, February 28, 1951, Dwight D. Eisenhower Presidential Library, Abilene, Kans.

43 **In 1991 he wrote:** Charles Solomon, "Two States—One Nation?," *Los Angeles Times*, November 17, 1991.

44 **Article 9 of the nation's constitution:** Constitution of Japan, ch. II, art. 9 (effective May 3, 1947).

44 **The total cost of the program:** Stephen Losey, "F-35s to Cost $2 Trillion as Pentagon Plans Longer Use, Says Watchdog," *Defense News*, April 15, 2024.

44 **But as General Mark Milley:** Mark A. Milley, "The Future of Geopolitics and the Role of Innovation and Technology," Washington D.C., May 7, 2024.

45 **The arrival of swarms of drones:** See Paul Mozur and Adam Satariano, "A.I. Begins Ushering in an Age of Killer Robots," *New York Times*, July 2, 2024.

45 **The U.S. Department of Defense:** Office of the Under Secretary of Defense, U.S. Department of Defense Fiscal Year 2024 Budget Request (March 2023); see also Will Henshall, "The U.S. Military's Investments into Artificial Intelligence Are Skyrocketing," *Time*, March 29, 2024.

47 **As David Graeber, who taught:** David Graeber, "Of Flying Cars and the Declining Rate of Profit," *Baffler*, no. 9 (March 2012): 77.

47 **After the company changed course:** Graeme Hanna, "'Stop Working with Pentagon'—OpenAI Staff Face Protests," *ReadWrite*, February 13, 2024; Ellen Huet, "Protesters Gather Outside OpenAI Headquarters," *Bloomberg*, February 13, 2024.

48 **In an essay:** Nicholas Negroponte, "Big Idea Famine," *Journal of Design and Science*, no. 3 (February 2018).

49 **As Gordon has written:** Robert J. Gordon, "The End of Economic Growth," *Prospect*, January 21, 2016.

49 **A profile of Musk:** Jill Lepore, "The X-Man," *New Yorker*, September 11, 2023.

49 **Musk's critics are often far:** Theodore Roosevelt, "Citizenship in a Republic:

Address Delivered at the Sorbonne, Paris, April 23, 1910," in *Presidential Addresses and State Papers and European Addresses: December 8, 1908, to June 7, 1910* (New York: Review of Reviews, 1910), 2191.

49 **For years, many were convinced:** Ashlee Vance, *Elon Musk: Tesla, SpaceX, and the Quest for a Fantastic Future* (New York: HarperCollins, 2015), 218.

50 **As Peter Thiel observed:** George Packer, "No Death, No Taxes," *New Yorker*, November 20, 2011.

50 **In 2022, YouTube made $959 million:** Amanda Raffoul et al., "Social Media Platforms Generate Billions of Dollars in Revenue from U.S. Youth: Finding from a Simulated Model," *PLoS One*, December 27, 2023.

50 **Let us not go gentle:** See Dylan Thomas, *The Collected Poems of Dylan Thomas: Original Edition* (New York: New Directions, 2010), 122.

51 **"He ate bitterness":** Barbara Demick and David Pierson, "China Political Star Xi Jinping a Study in Contrasts," *Los Angeles Times*, February 11, 2012.

51 **Xi's older sister:** Chris Buckley and Didi Kirsten Tatlow, "Cultural Revolution Shaped Xi Jinping, from Schoolboy to Survivor," *New York Times*, September 24, 2015.

51 **An official government account:** Buckley and Tatlow, "Cultural Revolution Shaped Xi"; Evan Osnos, "Xi Jinping's Historic Bid at the Communist Party Congress," *New Yorker*, October 23, 2022.

51 **As a professor:** Osnos, "Xi Jingping's Historic Bid."

51 **Anne Applebaum rightly reminds:** Anne Applebaum, "There Is No Liberal World Order," *Atlantic*, March 31, 2022.

52 **But as Henry Kissinger has observed:** Henry Kissinger, foreword to *From Third World to First: The Singapore Story: 1965–2000*, by Lee Kuan Yew (New York: HarperCollins, 2000), ix.

52 **In 1985, he spent:** Mike Wendling, "Xi Jinping: Chinese Leader's Surprising Ties to Rural Iowa," BBC, November 15, 2023.

52 **And Xi's only daughter:** Evan Osnos, "What Did China's First Daughter Find in America?," *New Yorker*, April 6, 2015.

52 **A reporter for a Japanese newspaper:** Osnos, "China's First Daughter."

52 **On a later trip, Xi mentioned:** "President Xi's Speech on China-US Ties," *China Daily*, September 24, 2015; see also Taisu Zhang, Graham Webster, and Orville Schell, "What Xi Jinping's Seattle Speech Might Mean for the U.S.," *Foreign Policy*, September 23, 2015; Austin Ramzy, "Xi Jinping on 'House of Cards' and Hemingway," *New York Times*, September 23, 2015.

53 **But as Chloé Morin:** Catherine Porter, "Cheers, Fears, and 'Le Wokisme': How the World Sees U.S. Campus Protests," *New York Times*, May 3, 2024.

53 **This "moral dualism":** Remi Adekoya, "The Oppressed vs. Oppressor Mistake," Institute of Art and Ideas, October 17, 2023.

53 **Yet we still cling:** Lawrence H. Keeley, *War Before Civilization: The Myth of the Peaceful Savage* (New York: Oxford University Press, 1996), 52, 102.

54 **"Never in history has violence":** Paulo Freire, *Pedagogy of the Oppressed*, trans. Myra Bergman Ramos (Penguin Books, 2017), 29 (cited by Adekoya).

CHAPTER FIVE:
THE ABANDONMENT OF BELIEF

57 **Neier was born:** Aryeh Neier, *Defending My Enemy: American Nazis, the Skokie Case, and the Risks of Freedom* (New York: E. P. Dutton, 1979), 2.

57 **He later estimated:** Tom Goldstein, "Neier Is Quitting Post at A.C.L.U.; He Denies Link to Defense of Nazis," *New York Times*, April 18, 1978.

58 **"To defend myself":** Neier, *Defending My Enemy*, 5.

58 **A decade before:** Peter Salovey, "Free Speech, Personified," *New York Times*, November 26, 2017.

58 **Earlier that year:** George C. Wallace, "Inaugural Address," Montgomery, Ala., January 14, 1963, Alabama Department of Archives and History.

58 **His potential arrival:** Efrem Sigel, "New Wallace Invitation Expected at Yale Today," *Harvard Crimson*, September 24, 1963.

58 **Pauli Murray, who was pursuing:** Salovey, "Free Speech, Personified"; Kathryn Schulz, "The Many Lives of Pauli Murray," *New Yorker*, April 10, 2017.

58 **Murray, born in Baltimore:** Schulz, "Many Lives of Pauli Murray."

59 **Her father had been committed:** Schulz, "Many Lives of Pauli Murray."

59 **Murray's maternal grandmother:** See Pauli Murray, *Song in a Weary Throat: Memoir of an American Pilgrimage* (New York: Liveright, 2018).

59 **She argued that even though:** Salovey, "Free Speech, Personified."

59 **Murray anticipated the risk:** Harry Kalven Jr., *The Negro and the First Amendment* (Columbus: Ohio State University Press, 1965); see also *Brown v. Louisiana*, 383 U.S. 131 (1966), n1 (citing Kalven).

59 **The Yale Political Union:** Efrem Sigel, "Harvard, Yale Students to Issue New Invitations to Gov. Wallace," *Harvard Crimson*, September 25, 1963; see also Sam Chauncey Jr., letter to the editor, *Yale Daily News*, November 29, 2017 (recounting that Brewster would later say that he "made the wrong decision from the point of view of principle, but did the right thing

when it came to preventing violence"); see also Nathaniel Zelinsky, "Challenging the Unchallengeable (Sort Of)," *Yale Alumni Magazine*, January/February 2015.

59 **In 2023, three university presidents:** Harriet Sherwood, "Hamas Says 250 People Held Hostage in Gaza," *Guardian*, October 16, 2023; Cassandra Vinograd and Isabel Kershner, "Israel's Attackers Took About 240 Hostages," *New York Times*, November 20, 2023.

60 **To many, the presidents were:** See, for example, Anemona Hartocollis, Stephanie Saul, and Vimal Patel, "At Harvard, a Battle Over What Should Be Said About the Hamas Attacks," *New York Times*, October 10, 2023.

60 **As Maureen Dowd pointed out:** Maureen Dowd, "The Ivy League Flunks Out," *New York Times*, December 9, 2023.

61 **But few seem interested:** See, for example, Matt Bai, *All the Truth Is Out: The Week Politics Went Tabloid* (New York: Vintage, 2014).

61 **It is "the proliferation of frenzies":** Daniel Sutter, "Media Scrutiny and the Quality of Public Officials," *Public Choice* 129 no. 1/2 (October 2006): 38; see also Larry J. Sabato, *Feeding Frenzy: How Attack Journalism Has Transformed American Politics* (New York: Free Press, 1993), 211 (arguing that "the price of power has been raised dramatically, far too high for many outstanding potential officeholders," and that "American society today is losing the services of many exceptionally talented individuals who could make outstanding contributions to the commonweal, but who understandably will not subject themselves and their loved ones to abusive, intrusive press coverage") (quoted in Andrew B. Hall, *Who Wants to Run? How the Devaluing of Political Office Drives Polarization* (Chicago: University of Chicago Press, 2019), 67).

61 **In 1991, Larry Sabato:** Sabato, *Feeding Frenzy* 4.

61 **Americans, in particular:** "Public Figures and Their Private Lives," *Time*, August 22, 1969.

62 **In 1952, Richard Nixon:** *American Experience: The Presidents*, "Nixon, Part One: The Quest," PBS, October 15, 1990.

62 **Her husband replied:** "Nixon, Part One," PBS.

63 **Appeals to virtue and character:** Garance Franke-Ruta, "Paul Harvey's 1978 'So God Made a Farmer' Speech," *Atlantic*, February 3, 2013.

64 **The shallow and thinly veiled:** Alistair Barr, "Google's 'Don't Be Evil' Becomes Alphabet's 'Do the Right Thing,'" *Wall Street Journal*, October 2, 2015.

64 **As the French author Pascal Bruckner:** Pascal Bruckner, *The Tears of the White Man: Compassion as Contempt*, trans. William R. Beer (New York:

Free Press, 1986), 69 (quoted by Roger Kimball, "The Perils of Designer Tribalism," *New Criterion*, April 2001).

64 **"If you do not feel it":** Johann Wolfgang von Goethe, *Faust*, trans. Abraham Hayward and A. Bucheim (London: George Bell and Sons, 1892), 40–41 (for a translation on which this one is based).

64 **We later learned that WilmerHale:** Lauren Hirsch, "One Law Firm Prepared Both Penn and Harvard for Hearing on Antisemitism," *New York Times*, December 8, 2023.

65 **But as Lawrence Summers:** Lawrence Summers, interview by David Remnick, *New Yorker Radio Hour*, NPR, May 3, 2024.

65 **In a collection of essays:** Erving Goffman, *Asylums: Essays on the Social Situation of Mental Patients and Other Inmates* (London: Taylor & Francis, 2017), xxi.

65 **A series of civil rights demonstrations:** There are conflicting accounts regarding the number of bombs that exploded that night. Compare Paul Bass and Douglas W. Rae, *Murder in the Model City: The Black Panthers, Yale, and the Redemption of a Killer* (New York: Basic Books, 2006), 159 (two bombs), with Geoffrey Kabaservice, *The Guardians: Kingman Brewster, His Circle, and the Rise of the Liberal Establishment* (New York: Henry Holt, 2004), 4 (three bombs).

65 **In April 1970:** Joseph B. Treaster, "Brewster Doubts Fair Black Trials," *New York Times*, April 25, 1970, 1.

65 **He had ventured into:** Spiro Agnew, "Address at the Florida Republican Dinner," Fort Lauderdale, Fla., April 28, 1970, Yale University Library, New Haven, Conn.

66 **As Ralph Waldo Emerson:** Michael H. Hoffheimer, *Justice Holmes and The Natural Law* (New York: Routledge, 2013), 38.

66 **Our commitment to "openness":** Allan Bloom, *The Closing of the American Mind* (New York: Simon & Schuster, 1987), 56.

66 **Perry Link, the former professor:** Perry Link, "China: The Anaconda in the Chandelier," *New York Review of Books*, April 11, 2002.

67 **One Soviet directive:** Arlen Viktorovich Blium and Donna M. Farina, "Forbidden Topics: Early Soviet Censorship Directives," *Book History* 1 (1998): 273.

67 **"If I give my name":** Nicholas Fandos, "In an Online World, a New Generation of Protesters Chooses Anonymity," *New York Times*, May 2, 2024.

67 **But is a belief:** To be sure, there is a place, under certain conditions, for anonymous speech. See *McIntyre v. Ohio Elections Commission*, 514 U.S. 334 (1995), 357 (concluding that "anonymous pamphleteering is not a

pernicious, fraudulent practice, but an honorable tradition of advocacy and dissent").

68 **And the consequences of this inability:** Michael Sandel, *Liberalism and the Limits of Justice*, 2nd ed. (Cambridge, U.K.: Cambridge University Press, 1998), 217.

68 **"Where political discourse lacks":** Sandel, *Liberalism and the Limits of Justice*, 217.

CHAPTER SIX:
TECHNOLOGICAL AGNOSTICS

69 **Amy Gutmann, who taught:** Amy Gutmann, "Democratic Citizenship," *Boston Review*, October 1, 1994.

70 **As Manuel Castells Oliván:** Samuel P. Huntington, "Dead Souls: The Denationalization of the American Elite," *National Interest*, no. 75 (Spring 2004), 8 (quoting Castells).

70 **Hernán Cortés, the Spanish governor:** Winston A. Reynolds, "The Burning Ships of Hernán Cortés," *Hispania* 42, no. 3 (September 1959): 318.

71 **President Eisenhower warned:** Dwight D. Eisenhower, "Text of the Address by President Eisenhower," January 17, 1961, Dwight D. Eisenhower Presidential Library, Abilene, Kans. (quoted in Adrienne LaFrance, "The Rise of Techno-Authoritarianism," *Atlantic*, January 30, 2024).

71 **Our current era of innovation:** See LaFrance, "Rise of Techno-Authoritarianism" (arguing "no more 'build it because we can'").

71 **"They just can't wrap":** Ben Child, "Mark Zuckerberg Rejects His Portrayal in The Social Network," *Guardian*, October 20, 2010.

72 **As Stephen L. Carter, a professor:** Stephen L. Carter, *The Culture of Disbelief* (New York: Basic Books, 1993), 24.

72 **Carter noted that the roots:** Carter, *Culture of Disbelief*, 28.

72 **In an essay titled "Obsessive Actions":** Sigmund Freud, "Obsessive Actions and Religious Practices," in *Readings in Ritual Studies*, ed. Ronald L. Grimes (Upper Saddle River, N.J.: Prentice Hall, 1996), 216.

72 **As the German physicist Max Planck:** Max Planck, *Scientific Autobiography and Other Papers*, trans. Frank Gaynor (New York: Philosophical Library, 1949), 33.

73 **The soul of the country:** See Pascal Bruckner, *The Tears of the White Man: Compassion as Contempt*, trans. William R. Beer (New York: Free Press, 1986), 69 (cited by Roger Kimball, "The Perils of Designer Tribalism," *New Criterion*, April 2001).

73 **As Fukuyama has written:** Francis Fukuyama, "Waltzing with (Leo) Strauss," *American Interest* 10, no. 4 (February 2015).
74 **In her book *The Entrepreneurial State*:** Mariana Mazzucato, *The Entrepreneurial State: Debunking Public vs. Private Sector Myths* (London: Anthem Press, 2013), 84.
75 **A survey conducted in 2023:** Hannah J. Martinez, "The Graduating Class of 2023 by the Numbers," *Harvard Crimson*, 2023.
75 **Only 6 percent of graduates:** Aden Barton, "How Harvard Careerism Killed the Classroom," *Harvard Crimson*, April 21, 2023.
76 **The number of graduating college seniors:** David Armitage et al., *The Teaching of the Arts and Humanities at Harvard College: Mapping the Future* (Cambridge, Mass.: Harvard University, 2013), 7.
76 **At the same time, enrollment:** National Student Clearinghouse, "Computer Science Has Highest Increase in Bachelor's Earners," May 27, 2024.
76 **As Henry Kissinger reminded us:** Paul Kennedy, *The Rise and Fall of the Great Powers: Economic Change and Military Conflict from 1500 to 2000* (New York: Random House, 1989), 407–8.
77 **The airplanes are made:** Marcus Weisgerber, "F-35 Production Set to Quadruple as Massive Factory Retools," *Defense One*, May 6, 2016.
77 **The parts include $100,000:** Robert Levinson, "The F-35's Global Supply Chain," *Businessweek*, September 1, 2011.
77 **The fifty most valuable:** Analysis based on a review of publicly available financial information as of September 2024.
77 **The value of the American technology sector:** Sergei Klebnikov, "U.S. Tech Stocks Are Now Worth More Than $9 Trillion, Eclipsing the Entire European Stock Market," *Forbes*, December 15, 2020.
78 **Those who bristle at descriptions:** Ferdinand Lundberg, *America's 60 Families* (New York: Vanguard Press, 1937), 3.
79 **"If an upper class degenerates":** E. Digby Baltzell, *The Protestant Establishment: Aristocracy and Caste in America* (New Haven, Conn.: Yale University Press, 1987), 8.
79 **Even then, secluded and almost assuredly:** George Orwell, *1984* (New York: Penguin, 2023), 110.
79 **In East Germany, the state security:** Burkhard Bilger, "Piecing Together the Secrets of the Stasi," *New Yorker*, May 27, 2024.
80 **As Heller writes:** Ágnes Heller, *Beyond Justice* (Oxford: Blackwell, 1987), 273.
81 **The postmodern disinclination:** See Steven Brill, *The Death of Truth: How Social Media and the Internet Gave Snake Oil Salesmen and Demagogues the*

Weapons They Needed to Destroy Trust and Polarize the World—and What We Can Do (New York: Knopf, 2024).

81 **"The problem arises":** Morris Berman, *The Twilight of American Culture* (New York: W. W. Norton, 2006), 52.

81 **As Murrow reminded us:** Charles Kuralt, "Edward R. Murrow," *The North Carolina Historical Review* 48, no. 2 (April 1971), 170.

CHAPTER SEVEN: A BALLOON CUT LOOSE

84 **The supporters of a core curriculum:** William H. McNeill, "Western Civ in World Politics: What We Mean by the West," *Orbis* 41, no. 4 (Fall 1997): 520.

84 **William McNeill, for example:** McNeill, "Western Civ," 520.

85 **Kwame Anthony Appiah:** Kwame Anthony Appiah, "There Is No Such Thing as Western Civilisation," *Guardian*, November 9, 2016.

85 **Huntington's count of "major civilizations":** Samuel P. Huntington, "The Clash of Civilizations?," *Foreign Affairs* 72, no. 3 (Summer 1993): 25.

86 **Where were the fault lines:** Huntington, "Clash of Civilizations?," 29 (discussing "the fault lines between civilizations").

86 **"There is not *a* history":** Gilbert Allardyce, "The Rise and Fall of the Western Civilization Course," *American Historical Review* 87, no. 3 (June 1982): 719 (quoting Cheyette).

86 **He was born in New York City:** John Baldwin et al., "Memoirs of Fellows and Corresponding Fellows of the Medieval Academy of America," *Speculum* 91, no. 3 (July 2016): 894.

86 **Cheyette's academic interests:** See Fredric L. Cheyette, *Ermengard of Narbonne and the World of the Troubadours* (Ithaca, N.Y.: Cornell University Press, 2004).

87 **He articulated the dominant critique:** Fredric L. Cheyette, "Beyond Western Civilization: Rebuilding the Survey," *History Teacher* 10, no. 4 (August 1977): 535.

87 **In some cases, as one observer:** Lewis W. Spitz, "Beyond Western Civilization: Rebuilding the Survey," *History Teacher* 10, no. 4 (August 1977): 517.

87 **At Stanford, for example:** Suzette Leith, "Civ: Enlightenment . . . or the Black Death?," *Stanford Daily*, May 17, 1966, 1.

87 **But in November 1968:** Herbert L. Packer et al., *The Study of Education at Stanford: Report to the University*, vol. 2 (Stanford University, Novem-

ber 1968), 9; see also Bruce Kimball, ed., *The Liberal Arts Tradition: A Documentary History* (Lanham, Md.: University Press of America, 2010), 464.

88 **The course at Stanford ended:** John W. Coffey, "State of Higher Education: Chaos," *Stanford Daily*, November 29, 1971, 2.

88 **As one historian noted:** Allardyce, "Rise and Fall," 720.

88 **"We have Plato":** Joseph Tussman, "The Collegiate Rite of Passage," in *Experiment and Innovation: New Directions in Education at the University of California* (July 1968), excerpted in Packer et al., *Study of Education at Stanford*, 93; see also Allardyce, "Rise and Fall," 724 (noting that "all curricula are essentially religious").

88 **As a member:** Allardyce, "Rise and Fall," 702 (quoting Lucy M. Salmon in Andrew C. McLaughlin et al., *The Study of History in Schools: Report to the American Historical Association* (New York: Macmillan, 1899), 194).

89 **The current conception:** Appiah, "No Such Thing"; see also Oswald Spengler, *Decline of the West* (New York: Oxford University Press, 1991); Northrop Frye, "'The Decline of the West' by Oswald Spengler," *Daedalus* 103, no. 1 (Winter 1974); "Patterns in Chaos," review of *The Decline of the West: Perspectives of World History*, by Oswald Spengler, *Time*, December 10, 1928.

89 **As Winston Churchill observed:** Niall Ferguson, *Civilization: The West and the Rest* (New York: Penguin Books, 2011), 98 (quoting Winston Churchill's address at the University of Bristol on July 2, 1938).

89 **Claude-Lévi Strauss, the French anthropologist:** Claude Lévi-Strauss, *Tristes Tropiques*, trans. John Weightman and Doreen Weightman (New York: Penguin Books, 2012), 326 (cited by Bruckner, *Tears of the White Man*, 100).

89 **Appiah, for example, has argued:** Appiah, "No Such Thing."

90 **Adam Shatz, the U.S. editor:** Adam Shatz, "'Orientalism,' Then and Now," *New York Review of Books*, May 20, 2019.

90 **As Shatz put it:** Shatz, "'Orientalism,' Then and Now."

90 **The author Pankaj Mishra:** Pankaj Mishra, "The Reorientations of Edward Said," *New Yorker*, April 19, 2021.

91 **Indeed, the book gave birth:** Mishra, "Reorientations of Edward Said."

91 **In his biography of Said:** Timothy Brennan, *Places of Mind: A Life of Edward Said* (New York: Farrar, Straus and Giroux, 2021), 220.

91 **As Said himself explained:** Edward Said, *Orientalism* (New York: Vintage Books, 1994), 332.

92 **As Said made clear:** Said, *Orientalism*, 7 (citing Denys Hay, *Europe: The Emergence of an Idea*, 2nd ed. (Edinburgh: Edinburgh University Press, 1968)).

92 **Some have read Said:** See Brennan, *Places of Mind*, 205 (noting that Said "went to great lengths in lectures on the eve of *Orientalism*'s publication to attack postmodernism").

92 **As Mishra has observed:** Mishra, "Reorientations of Edward Said" (noting that "the book's critique of Eurocentrism was in fact curiously Eurocentric").

93 **McNeill wrote in an essay:** McNeill, "Western Civ," 521.

93 **As the historian Niall Ferguson:** Ferguson, *Civilization*, 6.

94 **The ability to reckon:** Nate Silver, *On the Edge: The Art of Risking Everything* (New York: Penguin Press, 2024), 25.

95 **The Germans, he wrote:** Vannevar Bush, *Modern Arms and Free Men* (New York: Simon & Schuster, 1949), 53.

CHAPTER EIGHT: "FLAWED SYSTEMS"

97 **In January 1970, *Time* magazine:** "Man and Woman of the Year: The Middle Americans," *Time*, January 5, 1970.

97 **The division of the country:** See, for example, Richard Nixon, "Address to the Nation on the War in Vietnam," Washington, D.C., November 3, 1969 (appealing to "the great silent majority of my fellow Americans"); Matthew D. Lassiter, "Who Speaks for the Silent Majority?," *New York Times*, November 2, 2011; see also Rick Perlstein, *Nixonland: The Rise of a President and the Fracturing of America* (New York: Scribner, 2008), 447.

98 **Lee Felsenstein, for example:** Walter Isaacson, *The Innovators: How a Group of Hackers, Geniuses, and Geeks Created the Digital Revolution* (New York: Simon & Schuster, 2015), 266.

98 **Stewart Brand, co-founder:** Stewart Brand, "We Owe It All to the Hippies," *Time*, March 1, 1995 (quoted in Isaacson, *Innovators*, 268).

99 **Levy articulated the ethos:** Steven Levy, *Hackers: Heroes of the Computer Revolution* (Sebastopol, Calif.: O'Reilly, 2010), 29–30.

100 **Steve Jobs, in particular:** Walter Isaacson, *Steve Jobs* (New York: Simon & Schuster, 2013), 34.

100 **As an undergraduate at Reed College:** Isaacson, *Steve Jobs*, 41.

100 **When attempting to court:** Malcolm Gladwell, "The Tweaker," *New Yorker*, November 6, 2011.

101 **Indeed, Apple objected:** Ellen Nakashima and Reed Albergotti, "The FBI Wanted to Unlock the San Bernardino Shooter's iPhone. It Turned to a Little-Known Australian Firm," *Washington Post*, April 14, 2021.

102 **The Macintosh, by contrast:** Erik Sandberg-Diment, "Hardware Review: Apple Weighs in with Its Macintosh," *New York Times*, January 24, 1984, C3.

102 **An initial draft of the advertisement:** David Burnham, "The Computer, the Consumer and Privacy," *New York Times*, March 4, 1984, E8.

CHAPTER NINE:
LOST IN TOYLAND

103 **In 1996, Toby Lenk:** Jeff Jensen, "Toby Lenk," *Advertising Age*, June 1, 1998.

103 **At its height:** Michael Sokolove, "How to Lose $850 Million—and Not Really Care," *New York Times Magazine*, June 9, 2002.

104 **"We're losing money fast":** Jensen, "Toby Lenk."

104 **An estimated fifty thousand companies:** Brian McCullough, *How the Internet Happened: From Netscape to the iPhone* (New York: Liveright, 2018), 141.

104 **As a *Wall Street Journal* profile:** Jason Fry, "eToys Story," *Wall Street Journal*, July 12, 1999.

104 **In May 1999, in its S-1:** Form S-1, Amendment No. 1, eToys Inc., U.S. Securities and Exchange Commission, May 19, 1999 (cited in George Anders and Ann Grimes, "eToys' Shares Nearly Quadruple, Outstripping Rival Toys 'R' Us," *Wall Street Journal*, May 21, 1999).

104 **"There is all this talk":** Brent Goldfarb and David A. Kirsch, *Bubbles and Crashes: The Boom and Bust of Technological Innovation* (Stanford, Calif.: Stanford University Press, 2019), 137 (citing Heather Green, "The Great Yuletide Shakeout," *Businessweek*, November 1, 1999, 22).

105 **"I just feel like I'm designed":** *Before Sunset*, directed by Richard Linklater (2004; Burbank, Calif.: Warner Independent Pictures).

106 **"I'm a keen golfer":** Jamie Doward, "A Gift-Horse in the Mouse," *Guardian*, October 23, 1999.

106 **The term itself would eventually:** See Leigh Alexander, "Why It's Time to Retire 'Disruption,' Silicon Valley's Emptiest Buzzword," *Guardian*, January 11, 2016 (describing the term "disruption" as having "the aftertaste of a sucked battery"); Adrian Daub, "The Disruption Con: Why Big Tech's Favourite Buzzword Is Nonsense," *Guardian*, September 24, 2020.

107 **But as Peter Turchin has argued:** Peter Turchin, *End Times: Elites, Counter-Elites, and the Path of Political Disintegration* (New York: Penguin Press, 2023), 89.

107 **Talcott Parsons, the American sociologist:** Talcott Parsons, "Certain Primary Sources and Patterns of Aggression in the Social Structure of the Western World," chap. 14 in *Essays in Sociological Theory*, rev. ed. (Glencoe, Ill.: Free Press, 1954), 314.

108 **An essay in *Commentary* magazine:** "The Study of Man: On Talcott Parsons," *Commentary*, December 1962.

108 **In an essay on human aggression:** Parsons, "Patterns of Aggression," 314.

108 **In February 2001:** John Cassy, "eToys Files for Bankruptcy," *Guardian*, February 28, 2001.

109 **"A year ago Americans":** "The Dot-Com Bubble Bursts," *New York Times*, December 24, 2000.

109 **For his part, Lenk blamed:** Sokolove, "How to Lose $850 Million."

109 **In a postmortem of the crash:** D. Quinn Mills, "Who's To Blame for the Bubble?," *Harvard Business Review*, May 2001.

109 **The *Guardian* noted:** Jane Martinson and Larry Elliott, "The Year Dot.com Turned into Dot.bomb," *Guardian*, December 29, 2000.

110 **As David Graeber wrote:** David Graeber, "Of Flying Cars and the Declining Rate of Profit," *Baffler*, no. 19 (March 2012).

110 **For Graeber, who described himself:** David Graeber, "The New Anarchists," *New Left Review* 13 (January/February 2002).

110 **And the lasting effects and harm:** Peter Gray, "The Decline of Play and the Rise of Psychopathology in Children and Adolescents," *American Journal of Play* 3, no. 4 (2011).

110 **At a gathering of lobbyists:** Alan Greenspan, "Remarks by Chairman Alan Greenspan: At the Annual Dinner and Francis Boyer Lecture of the American Enterprise Institute for Public Policy Research," Washington, D.C., December 5, 1996.

110 **The remark has come:** Robert J. Shiller, *Irrational Exuberance* (New York: Crown, 2006).

CHAPTER TEN:
THE ECK SWARM

115 **On June 26, 1951:** Martin Lindauer, "House-Hunting by Honey Bee Swarms," trans. P. Kirk Visscher, Karin Behrens, and Susanne Kuehnholz, *Journal of Comparative Physiology* 37 (1955): 271.

115 **Lindauer was born in 1918:** Thomas D. Seeley, "Martin Lindauer (1918–2008)," *Nature*, December 11, 2008, 718.

- 115 **Thomas D. Seeley, a biology professor:** Thomas D. Seeley, *Honeybee Democracy* (Princeton, N.J.: Princeton University Press, 2010), 13 (citing T. D. Seeley, S. Kühnholz, and R. H. Seeley, "An Early Chapter in Behavioral Physiology and Sociobiology: The Science of Martin Lindauer," *Journal of Comparative Physiology*, 188 (July 2002): 442).
- 116 **When animals search:** Lindauer, "Honey Bee Swarms," 264.
- 116 **Whereas most animals explore:** Lindauer, "Honey Bee Swarms," 264 (emphasis added).
- 116 **The University of Munich's Zoological Institute:** Lindauer, "Honey Bee Swarms," 265.
- 117 **At around three that afternoon:** Lindauer, "Honey Bee Swarms," 271–72.
- 117 **The bees return:** Karl von Frisch, *The Dance Language and Orientation of Bees*, trans. Leigh E. Chadwick (Cambridge, Mass.: Belknap Press of Harvard University Press, 1967), 269–70.
- 117 **Over the course of the afternoon:** Lindauer, "Honey Bee Swarms," 272–73.
- 117 **It had become evident:** Lindauer, "Honey Bee Swarms," 265–66, 287.
- 119 **Lindauer narrated the scene:** Lindauer, "Honey Bee Swarms," 272.
- 119 **On the following day:** Lindauer, "Honey Bee Swarms," 274.
- 119 **A new batch of potential sites:** Lindauer, "Honey Bee Swarms," 275.
- 119 **Over the next several hours:** Lindauer, "Honey Bee Swarms," 275.
- 119 **The names of the swarms:** Lindauer, "Honey Bee Swarms," 268.
- 120 **As one group of researchers:** Sayra Cristancho and Graham Thompson, "Building Resilient Healthcare Teams: Insights from Analogy to the Social Biology of Ants, Honey Bees and Other Social Insects," *Perspectives on Medical Education* 12, no. 1 (2023): 254.
- 120 **Each of the cameras:** Giorgio Parisi, *In a Flight of Starlings: The Wonder of Complex Systems*, trans. Simon Carnell (New York: Penguin Books, 2023), 9.
- 120 **He found that the flocks:** Parisi, *Flight of Starlings*, 11.
- 121 **As Parisi wrote:** Parisi, *Flight of Starlings*, 16.

CHAPTER ELEVEN:
THE IMPROVISATIONAL STARTUP

- 122 **Johnstone is credited with articulating:** Theresa Robbins Dudeck, *Keith Johnstone: A Critical Biography* (London: Bloomsbury, 2013), 12.
- 122 **Jerry Seinfeld has said:** David Remnick, "The Scholar of Comedy," *New Yorker*, April 28, 2024.

123 **He was born in 1933:** Dudeck, *Keith Johnstone*, 20.
123 **One of his central insights:** Keith Johnstone, *Impro: Improvisation and the Theatre* (New York: Routledge, 1981), 41–52.
123 **Johnstone's interest and approach:** Dudeck, *Keith Johnstone*, 12.
124 **The most dominant jackdaws:** Konrad Z. Lorenz, *King Solomon's Ring* (New York: Thomas Y. Crowell, 2020), 149.
124 **For Johnstone, "every inflection":** Johnstone, *Impro*, 33.
124 **In particular, a bifurcation:** Johnstone, *Impro*, 36.
125 **By the 1960s:** *American Experience: Silicon Valley*, "Silicon Valley: Chapter 1," directed by Randall MacLowry, PBS, February 5, 2013.
125 **Every human institution:** According to one survey of three hundred large businesses in the United States, the average number of people reporting to a company's chief executive officer nearly doubled from the 1980s to the 1990s, increasing from approximately four people in 1986 to eight people more than a decade later, in 1998. See Raghuram Rajan and Julie Wulf, "The Flattening Firm: Evidence from Panel Data on the Changing Nature of Corporate Hierarchies," Working Paper No. 9633 (National Bureau of Economic Research, April 2003), 4.
126 **A group of researchers:** Leslie A. Perlow, Constance Noonan Hadley, and Eunice Eun, "Stop the Meeting Madness," *Harvard Business Review*, July–August 2017.
127 **A symphony orchestra:** Peter F. Drucker, "The Coming of the New Organization," *Harvard Business Review*, January 1988.

CHAPTER TWELVE:
THE DISAPPROVAL OF THE CROWD

130 **Asch was born in Warsaw:** John Ceraso, Irvin Rock, and Howard Gruber, "On Solomon Asch," in *The Legacy of Solomon Asch*, ed. Irvin Rock (Hillsdale, N.J.: Lawrence Erlbaum Associates, 1990), 3.
130 **When he was thirteen:** David Stout, "Solomon Asch Is Dead at 88; A Leading Social Psychologist," *New York Times*, February 29, 1996, D19.
130 **The other seven:** Solomon E. Asch, "Effects of Group Pressure upon the Modification and Distortion of Judgments," in *Groups, Leadership, and Men: Research in Human Relations*, ed. Harold Guetzkow (Pittsburgh: Carnegie Press, 1951), 178.
131 **As Asch later wrote:** Asch, "Effects of Group Pressure," 179.
131 **For Asch, and many others:** Solomon E. Asch, "Opinions and Social Pressure," *Scientific American* 193, no. 5 (November 1955): 34.

NOTES

131 **A friend and colleague:** Ceraso, Rock, and Gruber, "On Solomon Asch," 8.

132 **The institutional review boards:** Christine Grady, "Institutional Review Boards: Purpose and Challenges," *Chest* 148, no. 5 (November 2015): 1150.

132 **Milgram was born in 1933:** Thomas Blass, *The Man Who Shocked the World: The Life and Legacy of Stanley Milgram* (New York: Basic Books, 2009), 1.

133 **An advertisement seeking volunteers:** Stanley Milgram, *Obedience to Authority: An Experimental View* (New York: Harper Perennial, 2009), 14.

133 **Each of the volunteers:** Milgram, *Obedience to Authority*, 14.

133 **The test subjects were told:** Milgram, *Obedience to Authority*, 19.

133 **At the outset:** Milgram, *Obedience to Authority*, 20.

133 **The learner, of course:** Milgram, *Obedience to Authority*, 16.

133 **Of the dozens of individuals:** Milgram, *Obedience to Authority*, 5.

133 **The results captivated the country:** Walter Sullivan, "65% in Test Blindly Obey Order to Inflict Pain," *New York Times*, October 26, 1963, 10.

134 **In one of the most haunting:** Milgram, *Obedience to Authority*, 73.

134 **"Subject: I can't stand it":** Milgram, *Obedience to Authority*, 73.

134 **As Milgram put it:** Milgram, *Obedience to Authority*, 77.

135 **"He thinks he is killing someone":** Milgram, *Obedience to Authority*, 77.

135 **Milgram's experiment provided:** See Hannah Arendt, *Eichmann in Jerusalem: A Report on the Banality of Evil* (New York: Viking Press, 1963).

135 **Not all of Milgram's subjects:** Milgram, *Obedience to Authority*, 84.

135 **The investigator leading the session:** Milgram, *Obedience to Authority*, 16, 85.

135 **He also repeated:** Milgram, *Obedience to Authority*, 85.

135 **The psychological resilience:** Milgram, *Obedience to Authority*, 85.

136 **The group of experiments:** The experiment also serves as a reminder of how gentle contemporary review boards in psychology departments have grown, approving only the most mild forms of deception in experiments conducted on volunteers and perhaps forgoing entire lines of productive and valuable research into the human mind.

136 **As Howard Gruber:** Ceraso, Rock, and Gruber, "On Solomon Asch," 8.

137 **In September 1922:** *Monet's Years at Giverny* (New York: Metropolitan Museum of Art, 1978), 34–36.

137 **His later works:** Emily Watlington, "'Monet/Mitchell' Shows How the Impressionist's Blindness Charted a Path for Abstraction," *Art in America*, May 12, 2023.

137 **A retrospective of the painter's work:** See Suzanne Pagé, Marianne Mathieu, and Angeline Scherf, *Monet—Mitchell* (New Haven, Conn.: Yale University Press, 2022).

137 **Similarly, when Ludwig van Beethoven:** Matthew Guerrieri, *The First Four Notes: Beethoven's Fifth and the Human Imagination* (New York: Vintage Books, 2014), 8.

138 **As news, however:** Guerrieri, *First Four Notes*, 12.

138 **Some have argued:** Robin Wallace, "Why Beethoven's Loss of Hearing Added Dimensions to His Music," *Zócalo Public Square*, July 28, 2019.

CHAPTER THIRTEEN:
BUILDING A BETTER RIFLE

139 **On September 28, 2011:** Kevin Nevers, "'He Didn't Hesitate': Airborne Medic Jim Butz Dies a Hero in Afghanistan," *Chesterton* (Ind.) *Tribune*, October 3, 2011.

139 **The stretch of land:** Milton Bearden, "Afghanistan, Graveyard of Empires," *Foreign Affairs*, November 1, 2001.

139 **James Butz:** John Byrne, "Northwest Indiana Medic Killed in Afghanistan," *Chicago Tribune*, October 1, 2011; Susan Brown, "Soldier's Dad: 'He'll Always Be My Hero,'" *Times of Northwest Indiana*, October 2, 2011.

139 **A second explosion:** Brown, "Soldier's Dad."

139 **"Jimmy didn't hesitate":** Nevers, "'He Didn't Hesitate.'"

140 **By 2012, more than three thousand:** Harvey M. Sapolsky and Michael Schrage, "More Than Technology Needed to Defeat Roadside Bombs," *National Defense*, April 2012, 17.

140 **A total of 14,500:** Umar Farooq, "Pakistani Fertilizer Grows Both Taliban Bombs and Afghan Crops," *Christian Science Monitor*, May 9, 2013.

140 **As a U.S. Navy officer:** Jason Shell, "How the IED Won: Dispelling the Myth of Tactical Success and Innovation," *War on the Rocks*, May 1, 2017.

140 **The U.S. military spent:** Sapolsky and Schrage, "More Than Technology," 17; Shell, "How the IED Won" (estimating the cost of an IED at $265).

140 **The U.S. Army decided:** "Oshkosh MRAP All Terrain Vehicle," *Army Technology*, September 14, 2009.

140 **By October 2012:** Alex Rogers, "The MRAP: Brilliant Buy, or Billions Wasted?," *Time*, October 2, 2012.

140 **The more powerful explosive devices:** Sapolsky and Schrage, "More Than Technology," 17. Some also questioned whether the newer vehicles with more substantial armor even offered much more protection than existing personnel carriers. See Chris Rohlfs and Ryan Sullivan, "Why the $600,000 Vehicles Aren't Worth the Money," *Foreign Affairs*, July 26, 2012.

141 **The frustration of so many soldiers:** See Annie Jacobsen, "Palantir's God's-

Eye View of Afghanistan," *Wired*, January 20, 2021; Robert Draper, "Boondoggle Goes Boom," *New Republic*, June 18, 2013.

141 **In another era:** Arthur Herman, "What if Apple Designed an iFighter?," *Wall Street Journal*, July 23, 2012.

142 **An intelligence officer in Afghanistan:** "Army 'Rapid Equipping Force' Taking Root, Chief Says," *National Defense*, October 1, 2006.

142 **The officer wrote that:** Darrel Issa and Jason Chaffetz to Leon E. Panetta, August 1, 2012, United States House Committee on Oversight and Government Reform.

143 **In January 2012:** *Palantir Technologies v. United States*, No. 16-Civ-784-MBH (Fed. Claims, June 30, 2016), 49; see also Rowan Scarborough, "Soldier Battling Bombs Irked by Software Switch," *Washington Times*, July 22, 2012; Steven Brill, "Trump, Palantir, and the Battle to Clean Up a Huge Army Procurement Swamp," *Fortune*, March 27, 2017.

143 **"We aren't going to sit here":** Scarborough, "Soldier Battling Bombs."

143 **A deputy to James Mattis:** Brill, "Battle to Clean Up."

144 **Over twenty years:** U.S. Department of Defense, Casualty Status, July 16, 2024; "Costs of War: Afghan Civilians," Watson Institute of International and Public Affairs, Brown University, Providence, R.I.

144 **The conflict would end up costing:** Christopher Helman and Hank Tucker, "The War in Afghanistan Cost America $300 Million per Day for 20 Years, with Big Bills Yet to Come," *Forbes*, August 16, 2021.

144 **As of August 2006:** "Absence of America's Upper Classes from the Military," *ABC News*, August 3, 2006.

145 **If a battle abroad:** Leo Shane III, "Why One Lawmaker Keeps Pushing for a New Military Draft," *Military Times*, March 30, 2015.

145 **In the American system:** Patrick Caddell, interviewed in "Jimmy Carter," *American Experience*, PBS, November 11, 2002.

146 **The litany of requirements:** Brad Orton, "Remarks at the National Performance Review Press Conference," October 26, 1993, Old Executive Office Building, Washington, D.C., C-SPAN.

146 **But developing an alternative:** Orton, "Remarks."

147 **They reached out to:** Orton, "Remarks"; Stephen Barr, "Clinton Proposed Procurement Changes," *Washington Post*, October 27, 1993.

147 **Senator William Roth:** William Roth, S. 1587, Federal Acquisition Streamlining Act of 1993, Committee on Governmental Affairs and the Committee on Armed Services, Washington, D.C., February 24, 1994, 4.

147 **At the time, the U.S. government:** Al Gore, *Common Sense Government: Works Better and Costs Less* (1998), 74.

148 **One list from the 1980s:** U.S. Department of Defense, "Military Specification: Cookie Mix, Dry," MIL-C-43205G, 7.

148 **A commission established by:** Richard D. White Jr., "Executive Reorganization, Theodore Roosevelt, and the Keep Commission," *Administrative Theory and Praxis* 24, no. 3 (2002): 512.

148 **Gifford Pinchot:** White Jr., "Executive Reorganization," 511–12; Danny Freedman, "They're Getting Rid of 'Red Tape' in Washington. Literally," *Washington Post*, January 16, 2023 (discussing the roots of the term "red tape").

148 **In 1983, for example:** Wayne Biddle, "House Approves Stiff Rules to Control Cost of Military Spare Parts," *New York Times*, May 31, 1984, B24.

148 **Some of the prices:** James Fairhall, "The Case for the $435 Hammer," *Washington Monthly*, January 1, 1987.

149 **The hammers, for example:** Airon A. Mothershed, "The $435 Hammer and $600 Toilet Seat Scandals: Does Media Coverage of Procurement Scandals Lead to Procurement Reform?," *Public Contract Law* 41, no. 4 (Summer 2012): 861.

149 **In 1984, a journalist:** Brad Knickerbocker, "Pentagon Steps Up Its War on Unscrupulous Defense Contractors," *Christian Science Monitor*, March 15, 1984 (quoted in Mothershed, "$435 Hammer and $600 Toilet Seat Scandals," 863).

149 **He would later say:** William J. Clinton, "State of the Union Address," Washington, D.C., January 23, 1996.

149 **At a press conference:** William J. Clinton, "Remarks Announcing the Report of the National Performance Review and an Exchange with Reporters," Washington, D.C., September 7, 1993.

150 **David E. Rosenbaum:** David E. Rosenbaum, "Remaking Government: Few Disagree with Clinton's Overall Goal, but History Shows the Obstacles Ahead," *New York Times*, September 8, 1993, A1.

150 **Clinton had been working:** Barr, "Procurement Changes."

150 **"This should never happen":** William J. Clinton, "Remarks at the National Performance Review Press Conference," October 26, 1993, Old Executive Office Building, Washington, D.C., C-SPAN.

150 **Gore, who was standing by:** Al Gore, "Remarks at the National Performance Review Press Conference," October 26, 1993, Old Executive Office Building, Washington, D.C., C-SPAN.

150 **The prevailing regulatory regime:** Thomas J. Kelleher et al., *Smith, Currie, and Hancock's Federal Government Construction Contracts* (Hoboken, N.J.: Wiley, 2010), 89.

151 **At a Senate hearing:** Federal Acquisition Streamlining Act of 1993: Hearing on 1587, before the Committee on Governmental Affairs and the Committee on Armed Services, 103rd Cong., February 24, 1994, 2 (statement of John Glenn).

151 **As Glenn pointed out:** Federal Acquisition Streamlining Act of 1993, John Glenn, 2.

151 **The strategy of public servants:** Federal Acquisition Streamlining Act of 1993, John Glenn, 3.

151 **Steven Brill:** Brill, "Battle to Clean Up."

152 **"What will Jay Leno do?":** Clinton, "Remarks on Signing the Federal Acquisition Streamlining Act of 1994."

153 **This sort of litigation:** Lizette Chapman, "Inside Palantir's War with the U.S. Army," *Bloomberg*, October 28, 2016.

153 **The case came before:** *Palantir USG Inc. v. United States*, No. 16-784C (Fed. Claims, November 3, 2016), 97.

153 **In short, we had won:** A federal appellate court in Washington, D.C., upheld Judge Horn's ruling. *Palantir USG Inc. v. United States*, 904 F.3d 980 (Fed. Cir. 2018).

153 **John McCain:** Shane Harris, "Palantir Wins Competition to Build Army Intelligence System," *Washington Post*, March 26, 2019.

153 **A year later:** Harris, "Palantir Wins Competition."

153 **The U.S. military's turn:** Harris, "Palantir Wins Competition."

154 **Zynga, the video game maker:** Evelyn M. Rusli, "Zynga's Value, at $7 Billion, Is Milestone for Social Gaming," *New York Times*, December 15, 2011.

154 **At a valuation of $25 billion:** Nicole Perlroth, "The Groupon IPO: By the Numbers," *Forbes*, June 2, 2011.

154 **The company, which is still:** See Robert Channick, "Groupon Issues 'Going Concern' Warning as Chicago-Based Online Marketplace Terminates River North HQ Lease," *Chicago Tribune*, May 13, 2023; Eric J. Savitz, "Groupon Stock Craters. The Turnaround Is Taking Longer Than Hoped," *Barron's*, November 10, 2023.

CHAPTER FOURTEEN:
A CLOUD OR A CLOCK

156 **He taught:** Henry Adams, *Tom and Jack: The Intertwined Lives of Thomas Hart Benton and Jackson Pollock* (New York: Bloomsbury Press, 2009), 30.

156 **In an interview:** Pepe Karmel, ed., *Jackson Pollock: Interviews, Articles, and Reviews* (New York: Museum of Modern Art, 1999), 15; see also Erika

Doss, *Benton, Pollock, and the Politics of Modernism* (Chicago: University of Chicago Press, 1995), 330 (discussing interview).

156 **Benton initially thought little:** Thomas Hart Benton, *An Artist in America* (Columbia, Mo.: University of Missouri Press, 1968), 339 (cited in Emily Esfahani Smith, "The Friendship That Changed Art," *Artists Magazine* 35, no. 6 (July/August 2018)).

156 **As the comedian:** David Sims, "No, Really, I'm Awful," *Atlantic*, April 26, 2023.

157 **Richard Alan Friedman, a professor:** Jill Filipovic, "I Was Wrong About Trigger Warnings," *Atlantic*, August 9, 2023.

157 **The artist and the founder:** Jack Kerouac, *On the Road* (New York: Penguin Books, 1976), 5.

158 **"There is nothing special":** René Girard, "Generative Scapegoating," in *Violent Origins: Walter Burket, René Girard, and Jonathan Z. Smith on Ritual Killing and Cultural Formation*, ed. Robert G. Hammerton-Kelly (Stanford, Calif.: Stanford University Press, 1987), 123.

158 **For Ernst Kris, the Austrian psychoanalyst:** Ernst Kris, *Psychoanalytic Explorations in Art* (New York: International Universities, Press, 1952), 59.

159 **"For nonconformity":** Ralph Waldo Emerson, "Self-Reliance," in *Nature and Selected Essays*, ed. Larzer Ziff (Penguin Books, 2003), 123.

159 **But Emerson is right to ask:** Emerson, "Self-Reliance," 123–24.

160 **For Berlin, there was:** Isaiah Berlin, *The Hedgehog and the Fox* (London: Weidenfeld & Nicolson, 1954), 1.

160 **"The fox knows many things":** Berlin, *Hedgehog and the Fox*, 1.

160 **Herbert Hoover:** See Kenneth Whyte, *Hoover: An Extraordinary Life in Extraordinary Times* (New York: Alfred A. Knopf, 2017), 35, 68–69; Jeremy Mouat and Ian Phimister, "The Engineering of Herbert Hoover," *Pacific Historical Review* 77, no. 4 (November 2008): 555, 560.

160 **He wrote in his memoirs:** Herbert Hoover, *The Memoirs of Herbert Hoover: Years of Adventure, 1874–1920* (New York: Macmillan, 1953), 133.

161 **One must, as the American philosopher:** John Dewey, "Pragmatic America," in *America's Public Philosopher: Essays on Social Justice, Economics, Education, and the Future of Democracy*, ed. Eric Thomas Weber (New York: Columbia University Press, 2021), 52.

161 **Dewey took pride:** Dewey, "Pragmatic America," 51.

161 **A commitment to this sort:** Dewey, "Pragmatic America," 52.

161 **At least sixteen hundred German scientists:** Annie Jacobsen, *Operation Paperclip: The Secret Intelligence Program That Brought Nazi Scientists to America* (New York: Little, Brown, 2014), ix.

NOTES

161 **An officer in the U.S. Air Force:** Jacobsen, *Operation Paperclip*, 52.
162 **In his book:** Philip E. Tetlock, *Expert Political Judgment: How Good Is It? How Can We Know?* (Princeton, N.J.: Princeton University Press, 2005), 40; see also John Lewis Gaddis, *On Grand Strategy* (Penguin Books, 2019), 8–9 (discussing Tetlock).
162 **As Tetlock explained:** Tetlock, *Expert Political Judgment*, 40.
162 **Eugene Wigner:** Eugene Wigner, "The Unreasonable Effectiveness of Mathematics in the Natural Sciences," *Communications in Pure and Applied Mathematics* 13, no. 1 (February 1960), 2.
162 **But that same drive:** Tetlock, *Expert Political Judgment*, 40.
162 **He and his team:** Tetlock, *Expert Political Judgment*, 49, 254.
163 **Tetlock was interested:** Tetlock, *Expert Political Judgment*, 9.
163 **It turns out that:** Tetlock, *Expert Political Judgment* (methodological appendix).
164 **There are a number of ways:** Tetlock, *Expert Political Judgment*, 75n6.
164 **But he also posed:** Tetlock, *Expert Political Judgment*, 74.
164 **The "worst performers":** Tetlock, *Expert Political Judgment*, 80.
164 **In the late 1970s:** Ohno noted that the approach, of asking "why" five times, was built on the "habit of watching" that he learned from Sakichi Toyoda, whose son would go on to found Toyota Motor Corporation in the late 1930s. Taiichi Ohno, *Toyota Production System: Beyond Large-Scale Production* (Boca Raton, Fla.: CRC Press, 1988), 77–78.
164 **The approach, on its face:** Ohno, *Toyota Production System*, 17.
164 **In the context:** Ohno, *Toyota Production System*, 17.
164 **For Ohno who was born:** John Holusha, "Taiichi Ohno, Whose Car System Aided Toyota's Climb, Dies at 78," *New York Times*, May 31, 1990, D23.
164 **His father worked:** Holusha, "Taiichi Ohno," D23.
166 **As Lucian Freud:** Michael Auping, "Lucian Freud: The Last Interview," *Times* (London), January 28, 2012.
167 **Martin Gayford, a British art critic:** Martin Gayford, *Man with a Blue Scarf: On Sitting for a Portrait by Lucian Freud* (London: Thames & Hudson, 2019), 10.
167 **The artist once told:** Auping, "Lucian Freud."

CHAPTER FIFTEEN:
INTO THE DESERT

171 **Nearly eight hundred visitors:** Francis Galton, "Vox Populi," *Nature* 75, no. 1949 (March 1907): 450.

171	**It was a striking result:** See James Surowiecki, *The Wisdom of Crowds* (New York: Anchor Books, 2005).
171	**For Galton, the experiment:** Galton, "Vox Populi," 451.
172	**We have, as Michael Sandel:** Michael Sandel, *What Money Can't Buy: The Moral Limits of Markets* (New York: Farrar, Straus and Giroux, 2012), 12–13.
173	**An FBI file:** William J. Maxwell, ed. *James Baldwin: The FBI File* (New York: Arcade, 2017), 7.
173	**Such invasions of personal privacy:** Kenneth D. Ackerman, "Five Myths About J. Edgar Hoover," *Washington Post*, November 9, 2011.
173	**A number of defense contractors:** National Physical Laboratory, "Tracking People by Their 'Gait Signature,'" September 20, 2012.
174	**François-Marie Arouet:** Voltaire, *Zadig; or, The Book of Fate* (London, 1749), 53.
174	**In the eighteenth century:** William Blackstone, *Commentaries on the Laws of England in Four Books*, vol. 2 (Philadelphia: J. B. Lippincott, 1893), 587.
174	**Thomas Starkie:** Thomas Starkie, *A Practical Treatise on the Law of Evidence, and Digest of Proofs, in Civil and Criminal Proceedings*, vol. 1 (Boston: Wells & Lilly, 1826), 507.
174	**In 2012, Palantir began:** Ali Winston, "Palantir Has Secretly Been Using New Orleans to Test Its Predictive Policing Technology," *Verge*, February 27, 2018.
175	**The use of our platform:** Matt Sledge and Ramon Antonio Vargas, "Palantir's Crime-Fighting Software Causes Stir in New Orleans," *Times-Picayune*, March 1, 2018.
175	**In an essay published:** Jay Stanley, "New Orleans Program Offers Lessons in Pitfalls of Predictive Policing," American Civil Liberties Union, March 15, 2018.
175	**In June 2020:** Jay Greene, "Amazon Bans Police Use of Its Facial-Recognition Technology for a Year," *Washington Post*, June 10, 2020; Drew Harwell, "Amazon Extends Ban on Police Use of Its Facial Recognition Technology Indefinitely," *Washington Post*, May 18, 2021.
176	**That same month:** Bobby Allyn, "IBM Abandons Facial Recognition Products, Condemns Racially Biased Surveillance," NPR, June 9, 2020.
176	**The company's chief executive:** Jay Peters, "IBM Will No Longer Offer, Develop, or Research Facial Recognition Technology," *Verge*, June 8, 2020.
177	**The view that advanced technology:** Rob Henderson, "'Luxury Beliefs' Are

the Latest Status Symbol for Rich Americans," *New York Post*, August 17, 2019.

177 **Such beliefs:** David Brooks, "The Sins of the Educated Class," *New York Times*, June 6, 2024; see also Rob Henderson, *Troubled: A Memoir of Foster Care, Family, and Social Class* (New York: Gallery Books, 2024).

177 **When Peggy Noonan noted:** Peggy Noonan, "How Trump Lost Half of Washington," *Wall Street Journal*, April 25, 2019.

CHAPTER SIXTEEN:
PIETY AND ITS PRICE

179 **At one point, Rubenstein:** Jerome Powell, "The Honorable Jerome H. Powell," interview by David M. Rubenstein, Economic Club of Washington, D.C., February 7, 2023; see also Matthew Impelli, "Jerome Powell Salary Admission Sparks Debate," *Newsweek*, February 7, 2023.

180 **His compensation:** Heather Long, "Who Is Jerome Powell, Trump's Pick for the Nation's Most Powerful Economic Position?," *Washington Post*, November 2, 2017.

180 **The unintended consequence:** Daniel Krcmaric, Stephen C. Nelson, and Andrew Roberts, "Billionaire Politicians: A Global Perspective," *Perspectives on Politics*, October 25, 2023; see also Andrew B. Hall, *Who Wants to Run? How the Devaluing of Political Office Drives Polarization* (Chicago: University of Chicago Press, 2019), 70 (noting that "one probable result of the lessened salaries for legislators is that, by and large, only wealthy people will run for office").

180 **Members of the U.S. House:** Ida A. Brudnick, "Congressional Salaries and Allowances: In Brief," *Congressional Research Service*, June 27, 2024, 1.

181 **As Matthew Yglesias:** Matthew Yglesias, "Pay Congress More," *Vox*, May 10, 2019.

181 **He argued:** James Madison, *The Writings of James Madison*, vol. 3, ed. Gaillard Hunt (New York: G. P. Putnam's Sons, 1902), 253.

182 **By 2007:** Seth Mydans, "Singapore Announces 60 Percent Pay Raise for Ministers," *New York Times*, April 9, 2007.

182 **At a parliamentary debate:** Lee Kuan Yew, "In His Own Words: Higher Pay Will Attract Most Talented Team, So Country Can Prosper," *Straits Times*, November 1, 1994.

183 **Everyone present:** E. E. Kintner, "Admiral Rickover's Gamble," *Atlantic*, January 1959.

183 **On that evening:** Kintner, "Admiral Rickover's Gamble."
184 **In May 1955:** Richard G. Hewlett and Francis Duncan, *Nuclear Navy, 1946–1962* (Chicago: University of Chicago Press, 1974), 222.
184 **A U.S. Navy report:** Hewlett and Duncan, *Nuclear Navy*, 222.
184 **The plan to construct:** Marc Wortman, *Admiral Hyman Rickover: Engineer of Power* (New Haven, Conn.: Yale University Press, 2022), 4.
184 **His father, who was a tailor:** Thomas B. Allen and Norman Polmar, *Rickover: Father of the Nuclear Navy* (Washington, D.C.: Potomac Books, 2007), 7.
184 **The speed with which:** Kintner, "Admiral Rickover's Gamble."
184 **On several occasions:** Norman Polmar and Thomas B. Allen, *Rickover: Controversy and Genius: A Biography* (New York: Simon & Schuster, 1982), 272 (noting that such stories, "usually anonymous," were "difficult to verify thoroughly").
185 **When a deputy arrived:** Hyman G. Rickover, interview by Diane Sawyer, *60 Minutes*, CBS, 1984.
185 **But his reverence:** John W. Finney, "Rickover, Father of Nuclear Navy, Dies at 86," *New York Times*, July 9, 1986, A1.
185 **A report in 1985:** Wayne Biddle, "Navy Lists General Dynamics' Gifts to Rickover," *New York Times*, June 5, 1985, D7.
185 **Rickover admitted:** Wayne Biddle, "Rickover Tells Lehman He Gave Away Gifts," *New York Times*, June 11, 1985, D1.
186 **Rickover would later argue:** Biddle, "Rickover Tells Lehman," D23.
186 **The U.S. Navy concluded:** Wayne Biddle, "General Dynamics Draws Penalties on Navy Dealings," *New York Times*, May 22, 1985, A1.
186 **John Lehman:** Biddle, "General Dynamics Draws Penalties," A1.
186 **An editorial:** "Admiral Rickover and the Trinkets," *New York Times*, May 24, 1985, A24.
186 **William Proxmire:** "Admiral Rickover and the Trinkets," *New York Times*, A24.
186 **An obituary in *Time* magazine:** Michael Duffy, "Hyman George Rickover, 1900–1986: They Broke the Mold," *Time*, July 21, 1986.
188 **The hope that a governing class:** Plato, *The Republic*, trans. Desmond Lee (Penguin Books, 2007), 29.
188 **In *Permanence and Change*:** Kenneth Burke, *Permanence and Change* (University of California Press, 1984), 16.
188 **Florentius, after attempting to kill Benedict:** Gregory I, *The Life of Our Most Holy Father S. Benedict* (Rome: 1895), 37.
189 **But when an apprentice:** Gregory I, *Life of Our Most Holy Father*, 37.

CHAPTER SEVENTEEN:
THE NEXT THOUSAND YEARS

190 **He surveyed the size:** Robin Dunbar, "Co-Evolution of Neocortex Size, Group Size, and Language in Humans," *Behavioral and Brain Sciences* 16, no. 4 (1993); see also Yuval Noah Harari, *Sapiens: A Brief History of Humankind* (New York: HarperCollins, 2015).

190 **The Hutterites, for example:** U.S. Department of the Interior, National Register of Historic Places Inventory, *Historic Hutterite Colonies Thematic Resources*, 1979; Dunbar, "Co-Evolution of Neocortex Size" (citing Garrett Hardin, "Common Failing," *New Scientist* 102 (1988): 76).

190 **Similarly, a study:** F. Carlene Bryant, *We're All Kin: A Cultural Study of a Mountain Neighborhood* (Knoxville: University of Tennessee Press, 1981), 3–4 (cited by Dunbar).

190 **Dunbar, who was born in Liverpool:** Dunbar, "Co-Evolution of Neocortex Size," 688.

191 **The monkeys and great apes:** Dunbar, "Co-Evolution of Neocortex Size," 682.

191 **For humans, language:** See Benedict Anderson, *Imagined Communities: Reflections on the Origin and Spread of Nationalism*, rev. ed. (London: Verso, 2016).

191 **Without those "imagined linkage[s]":** Anderson, *Imagined Communities*, 33.

191 **In 2017, Emmanuel Macron:** Eugénie Bastié, "Emmanuel Macron, de la négation de la culture française à l'exaltation de la France éternelle," *Le Figaro*, June 5, 2023.

192 **Yves Jégo:** Yves Jégo, "Emmanuel Macron et le reniement de la culture française," *Le Figaro*, February 6, 2017.

193 **In June 1996:** "Le Pen Scores Own Goal with Team Slur," *Irish Times*, June 25, 1996.

194 **Indeed, the editors of the textbook:** Anders Breidlid, Fredrik Chr. Brøgger, Oyvind T. Gulliksen, and Torbjorn Sirevag, eds., *American Culture: An Anthology*, 2nd ed. (New York: Routledge, 2008), 3.

194 **"I am calculating":** Lee Kuan Yew, "Speech at the 28th Anniversary of Liquor Retailers' Association," Chinese Chamber of Commerce, Singapore, October 3, 1965, National Archives of Singapore.

195 **"We sang different songs":** Lee Kuan Yew, "Speech at the National Day Rally," Kallang Theatre, Singapore, August 17, 1986, National Archives of Singapore.

195 **For most of the twentieth century:** Patrick Chin Leong Ng, "A Study of

Attitudes Towards the Speak Mandarin Campaign in Singapore," *Intercultural Communication Studies* 23, no. 3 (2014): 54.

195 **The British colony:** John Newman, "Singapore's 'Speak Mandarin Campaign': The Educational Argument," *Southeast Asian Journal of Social Science* 14, no. 2 (1986): 53.

195 **A government review:** Goh Keng Swee, *Report on the Ministry of Education, Singapore*, 1-1. February 10, 1979.

195 **The authors of the report:** Goh, *Report on the Ministry of Education*, 1-10.

196 **"Singapore used to be":** Ian Johnson, "In Singapore, Chinese Dialects Revive After Decades of Restrictions," *New York Times*, August 26, 2017.

196 **For his part:** Lee Kuan Yew, "Speech at the Opening of the Speak Mandarin Campaign," Singapore Conference Hall, Singapore, September 21, 1984, National Archives of Singapore.

196 **Saravanan Gopinathan:** Saravanan Gopinathan, "Singapore's Language Policies: Strategies for a Plural Society," *Southeast Asian Affairs* (1979): 291.

196 **"This is a new phase":** Lee Kuan Yew, "Speech at National Day Rally."

197 **By 2023, its GDP:** World Bank Group, GDP Per Capita, Singapore, 2023.

197 **As Henry Kissinger put it:** Henry Kissinger, foreword to *From Third World to First: The Singapore Story: 1965–2000*, by Lee Kuan Yew (New York: HarperCollins, 2000), x.

197 **That ancient argument:** Thomas Carlyle, *On Heroes: Hero-Worship, and the Heroic in History* (London: James Fraser, 1841), 12.

197 **The Panthéon in Paris:** C. B. Black, *Paris and Excursions from Paris* (London: Sampson Low, Marston, Low & Searle, 1873), 45.

198 **Indeed, a generation:** See Anderson, *Imagined Communities*.

198 **Richard Sennett:** Richard Sennett, "The Identity Myth," *New York Times*, January 30, 1994, E17 (quoted by Roger Kimball, "Institutionalizing Our Demise: America vs. Multiculturalism," *New Criterion*, June 2004).

198 **The political philosopher:** Martha Nussbaum, "Patriotism and Cosmopolitanism," *Boston Review*, October 1, 1994 (quoted by Kimball, "Institutionalizing Our Demise").

199 **In 1882, Ernest Renan:** See "Reminiscences of Ernest Renan," *Atlantic*, August 1883 (noting that Renan described his ancestors in Brittany as "simple tillers of the earth and fishers of the sea").

199 **He was among the first writers:** Ernest Renan, *What Is a Nation? And Other Political Writings*, trans. and ed. M. F. N. Giglioli (New York: Columbia University Press, 2018), 247, 261.

199 **Palantir takes its name:** Jack Butler, "Does the Left Really Want to Argue

That Enjoying *Lord of the Rings* Is 'Far-Right'?," *National Review*, July 19, 2024.

200 **James K. A. Smith:** James K. A. Smith, "Reconsidering 'Civil Religion,'" *Comment*, May 11, 2017.

200 **It is true:** Robert Bellah, "Civil Religion in America," *Daedalus* 96, no. 1 (Winter 1967): 1, 18.

201 **It is the "pluralism":** Alasdair MacIntyre, *After Virtue: A Study in Moral Theory*, 3rd ed. (Notre Dame, Ind.: University of Notre Dame Press, 2007), 226.

201 **It is now time:** MacIntyre, *After Virtue*, 263.

201 **We must instead now conjure:** Kimball, "Institutionalizing Our Demise."

201 **Walser was born in 1927:** Thomas Kovach and Martin Walser, *The Burden of the Past: Martin Walser on Modern German Identity: Texts, Contexts, Commentary* (Rochester, N.Y.: Camden House, 2008), 2.

202 **His parents were Catholic:** Andreas Illmer, "German Writer Martin Walser Dies Aged 96," *Deutsche Welle*, July 28, 2023.

202 **It would later emerge:** "Dieter Hildebrandt soll in NSDAP gewesen sein," *Die Welt*, June 30, 2007.

202 **Walser told the magazine:** "Dieter Hildebrandt soll in NSDAP gewesen sein," *Die Welt* (noting also that Hans-Dieter Kreikamp, an official at the German archives, disputed Walser's account of being involuntarily enrolled in the Nazi Party, noting that a handwritten signature would have been required for membership at the time).

202 **Walser said:** Kovach and Walser, *Burden of the Past*, 89.

203 **Walser denounced efforts:** Kovach and Walser, *Burden of the Past*, 90–91.

203 **A commentator at the time:** Kathrin Schödel, "Normalising Cultural Memory? The 'Walser-Bubis Debate' and Martin Walser's Novel *Ein springender Brunnen*," in *Recasting German Identity: Culture, Politics, and Literature in the Berlin Republic*, ed. Stuart Taberner and Frank Finlay (Rochester, N.Y.: Camden House, 2002), 67.

203 **The audience during Walser's speech:** Amir Eshel, *Jewish Memories, German Futures: Recent Debates in Germany About the Past* (Bloomington, Ind.: Indiana University, 2001), 9.

203 **The moment was deeply cathartic:** David A. Kamenetzky, "The Debate on National Identity and the Martin Walser Speech: How Does Germany Reckon with Its Past?," *SAIS Review* 19, no. 2 (Summer–Fall 1999): 258.

203 **The day after the speech:** "Martin Walser bereut Verhalten gegenüber Ignatz Bubis," *Spiegel*, March 16, 2007; see also "Geistige Brandstiftung.

Bubis wendet sich gegen Walser," *Frankfurter Allgemeine*, October 13, 1998.

CHAPTER EIGHTEEN:
AN AESTHETIC POINT OF VIEW

205 **More than two million:** Gareth Harris, "Mary Beard BBC Segment on Kenneth Clark's Civilisation Renews Debate About Its Eurocentricity," *Art Newspaper*, April 29, 2024.

205 **Church services were rescheduled:** David Olusoga, "Civilisation Revisited," *Guardian*, February 3, 2018.

205 **The painting of sixteenth-century Rome:** Kenneth Clark, *Civilisation* (New York: Harper & Row, 1969), 174; see also Charles Rosen, review of *Civilisation*, by Kenneth Clark, *New York Review of Books*, May 7, 1970 (quoting Clark).

205 **He compared an African mask:** Clark, *Civilisation*, 2.

205 **Elsewhere he declined:** Clark, *Civilisation*, xvii.

206 **Mary Beard, a British author:** Mary Beard, "Kenneth Clark by James Stourton Review—Mary Beard on Civilisation Without Women," *Guardian*, October 1, 2016.

206 **Even modest attempts:** Peggy Noonan, "The Uglyfication of Everything," *Wall Street Journal*, May 2, 2024.

206 **As David Denby wrote:** David Denby, "In Darwin's Wake," *New Yorker*, July 21, 1997, 59 (cited in Morris Berman, *The Twilight of American Culture* (New York: W. W. Norton, 2006), 57).

206 **Thorstein Veblen:** Thorstein Veblen, *The Theory of the Leisure Class*, ed. Martha Banta (New York: Oxford University Press, 2007), 92.

208 **One art historian:** Boria Sax, *Imaginary Animals: The Monstrous, the Wondrous and the Human* (London: Reaktion Books, 2013), 94.

209 **When Odysseus asked:** *The Odyssey of Homer*, trans. George Hebert Palmer (Cambridge, Mass.: Houghton Mifflin, 1949), 185.

209 **The outperformance:** See, for example, Rüdiger Fahlenbrach, "Founder-CEOs, Investment Decisions, and Stock Market Performance," *Journal of Financial and Quantitative Analysis* 44, no. 2 (April 2009).

209 **He found that an investment approach:** Fahlenbrach, "Founder-CEOs," 440.

210 **He concluded that:** Fahlenbrach, "Founder-CEOs," 463.

211 **The team at Purdue:** Joon Mahn Lee, Jongsoo Kim, and Joonhyung Bae, "Founder CEOs and Innovation: Evidence from S&P 500 Firms," SSRN, February 17, 2016, 4.

211 **For Swensen:** David Swensen, "A Conversation with David Swensen," interview by Robert E. Rubin, Council on Foreign Relations, November 14, 2017.

212 **The early participants shared:** Akhil Reed Amar, *America's Constitution: A Biography* (New York: Random House, 2005), 275.

212 **She described working toward:** Ruth Benedict, *Patterns of Culture* (Boston: Houghton Mifflin, 2005), 278.

212 **For several generations of anthropologists:** Charles King, *Gods of the Upper Air: How a Circle of Renegade Anthropologists Reinvented Race, Sex, and Gender in the Twentieth Century* (New York: Anchor Books, 2020), 212–13.

213 **The conceit of this era:** King, *Gods of the Upper Air*, 212–13.

213 **The effective altruism movement:** See Peter Singer, *Animal Liberation: A New Ethics for Our Treatment of Animals* (New York: New York Review, 1975).

213 **But this approach provided:** Roger Scruton, "Animal Rights," *City Journal* (Summer 2000).

214 **Leo Strauss:** Leo Strauss, *What Is Political Philosophy?* (Chicago: University of Chicago Press, 1959), 18.

214 **He was also early:** Strauss, *What Is Political Philosophy?*, 21.

214 **But it was this "indifference":** Strauss, *What Is Political Philosophy?*, 18–19.

215 **For Lee Kuan Yew:** Roger T. Ames and Henry Rosemont Jr., *The Analects of Confucius: A Philosophical Translation* (New York: Ballantine Books, 1998), 60.

215 **This was someone who is:** Lee Kuan Yew, *Lee Kuan Yew: The Grand Master's Insights on China, the United States, and the World*, ed. Graham Allison and Robert D. Blackwill (Cambridge, Mass.: MIT Press, 2020), 131.

215 **Sallust wrote that:** Sallust, *The War with Catiline*, trans. J. C. Rolfe and rev. by John T. Ramsey, Loeb Classical Library (Cambridge, Mass.: Harvard University Press, 2013), 39.

215 **As Irving Kristol:** Irving Kristol, "Countercultures," *Commentary*, December 1994 (quoted by Roger Kimball, "Institutionalizing Our Demise: America vs. Multiculturalism," *New Criterion*, June 2004).

216 **John Rawls contended:** John Rawls, *Political Liberalism* (New York: Columbia University Press, 1993), 194.

216 **The old means:** See also E. D. Hirsch Jr., *Cultural Literacy: What Every American Needs to Know* (New York: Vintage, 1988).

Bibliography

ABC News. "Absence of America's Upper Classes from the Military." August 3, 2006.

Ackerman, Kenneth D. "Five Myths About J. Edgar Hoover." *Washington Post*, November 9, 2011.

Adams, Henry. *Tom and Jack: The Intertwined Lives of Thomas Hart Benton and Jackson Pollock.* New York: Bloomsbury Press, 2009.

Adekoya, Remi. "The Oppressed vs. Oppressor Mistake." Institute of Art and Ideas, October 17, 2023.

Agnew, Spiro. "Address at the Florida Republican Dinner." Fort Lauderdale, Fla., April 28, 1970, Yale University Library, New Haven, Conn.

Alexander, Leigh. "Why It's Time to Retire 'Disruption,' Silicon Valley's Emptiest Buzzword." *Guardian*, January 11, 2016.

Allardyce, Gilbert. "The Rise and Fall of the Western Civilization Course." *American Historical Review* 87, no. 3 (June 1982): 695–725.

Allen, Thomas B., and Norman Polmar. *Rickover: Father of the Nuclear Navy.* Washington, D.C.: Potomac Books, 2007.

Allyn, Bobby. "IBM Abandons Facial Recognition Products, Condemns Racially Biased Surveillance." NPR, June 9, 2020.

Amar, Akhil Reed. *America's Constitution: A Biography.* New York: Random House, 2005.

American Experience: The Presidents. "Nixon, Part One: The Quest." PBS, October 15, 1990.

American Experience: Silicon Valley. "Silicon Valley: Chapter 1." Directed by Randall MacLowry. PBS, February 5, 2013.

Ames, Roger T. and Henry Rosemont Jr. *The Analects of Confucius: A Philosophical Translation.* New York: Ballantine Books, 1998.

Anders, George and Ann Grimes. "eToys' Shares Nearly Quadruple, Outstripping Rival Toys 'R' Us." *Wall Street Journal*, May 21, 1999.

Andersen, Ross. "The Panopticon Is Already Here." *Atlantic*, September 2020.

Anderson, Benedict. *Imagined Communities: Reflections on the Origin and Spread of Nationalism.* Rev. ed. London: Verso, 2016.

Appiah, Kwame Anthony. "There Is No Such Thing as Western Civilisation." *Guardian*, November 9, 2016.

Applebaum, Anne. "There Is No Liberal World Order." *Atlantic*, March 31, 2022.

Arendt, Hannah. *Eichmann in Jerusalem: A Report on the Banality of Evil.* New York: Viking Press, 1963.

Armitage, David et al. *The Teaching of the Arts and Humanities at Harvard College: Mapping the Future.* Cambridge, Mass.: Harvard University, 2013.

Army Technology. "Oshkosh MRAP All Terrain Vehicle." September 14, 2009.

Asch, Solomon E. "Effects of Group Pressure upon the Modification and Distortion of Judgments." In *Groups, Leadership, and Men: Research in Human Relations*, edited by Harold Guetzko. Pittsburgh: Carnegie Press, 1951.

———. "Opinions and Social Pressure." *Scientific American* 193, no. 5 (November 1955): 3, 31–35.

Atlantic. "The Reminiscences of Ernest Renan." August 1883.

Auping, Michael. "Lucian Freud: The Last Interview." *Times* (London), January 28, 2012.

The Babylonian Talmud. Translated by Michael L. Rodkinson. Boston: Talmud Society, 1918.

Bai, Matt. *All the Truth Is Out: The Week Politics Went Tabloid.* New York: Vintage, 2014.

Baldwin, John, et al. "Memoirs of Fellows and Corresponding Fellows of the Medieval Academy of America." *Speculum* 91, no. 3 (July 2016): 889–907.

Baltzell, E. Digby. *The Protestant Establishment: Aristocracy and Caste in America.* New Haven, Conn.: Yale University Press, 1987.

Barnett, Lincoln. "J. Robert Oppenheimer." *Life*, October 10, 1949.

Barr, Alistair. "Google's 'Don't Be Evil' Becomes Alphabet's 'Do the Right Thing.'" *Wall Street Journal*, October 2, 2015.

Barr, Stephen. "Clinton Proposes Procurement Changes." *Washington Post*, October 27, 1993.

Barton, Aden. "How Harvard Careerism Killed the Classroom." *Harvard Crimson*, April 21, 2023.

Bass, Paul, and Douglas W. Rae. *Murder in the Model City: The Black Panthers, Yale, and the Redemption of a Killer.* New York: Basic Books, 2006.

Bastié, Eugénie. "Emmanuel Macron, de la négation de la culture française à l'exaltation de la France éternelle." *Le Figaro*, June 5, 2023.

Beard, Mary. "Kenneth Clark by James Stourton Review—Mary Beard on Civilisation Without Women." *Guardian*, October 1, 2016.

Bearden, Milton. "Afghanistan, Graveyard of Empires." *Foreign Affairs*, November 1, 2001.

Bellah, Robert. "Civil Religion in America." *Daedalus* 96, no. 1 (Winter 1967).

Bender, Emily M., et al. "On the Dangers of Stochastic Parrots: Can Language Models Be Too Big?" *Proceedings of the 2021 ACM Conference on Fairness, Accountability, and Transparency* (2021): 610–23.

Benedict, Ruth. *Patterns of Culture*. Boston: Houghton Mifflin, 2005.

Benton, Thomas Hart. *An Artist in America*. Columbia, Mo.: University of Missouri Press, 1968.

Bergman, Ronen. *Rise and Kill First: The Secret History of Israel's Targeted Assassinations*. New York: Random House, 2018.

Berlin, Isaiah. *The Hedgehog and the Fox*. London: Weidenfeld & Nicolson, 1954.

Berman, Morris. *The Twilight of American Culture*. New York: W. W. Norton, 2006.

Bernstein, Jeremy. "Oppenheimer's Beginnings." *New England Review* 25, no. 1/2 (Winter/Spring 2004): 38–51.

Biddle, Wayne. "General Dynamics Draws Penalties on Navy Dealings." *New York Times*, May 22, 1985.

———. "House Approves Stiff Rules to Control Cost of Military Spare Parts." *New York Times*, May 31, 1984.

———. "Navy Lists General Dynamics' Gifts to Rickover." *New York Times*, June 5, 1985.

———. "Rickover Tells Lehman He Gave Away Gifts." *New York Times*, June 11, 1985.

Bilger, Burkhard. "Piecing Together the Secrets of the Stasi." *New Yorker*, May 27, 2024.

Bird, Kai, and Martin J. Sherwin. *American Prometheus: The Triumph and Tragedy of J. Robert Oppenheimer*. New York: Alfred A. Knopf, 2005.

Black, C. B. *Paris and Excursions from Paris*. London: Sampson Low, Marston, Low & Searle, 1873.

Blackstone, William. *Commentaries on the Laws of England in Four Books*, Vol. 2. Philadelphia: J. B. Lippincott, 1893.

Blass, Thomas. *The Man Who Shocked the World: The Life and Legacy of Stanley Milgram*. New York: Basic Books, 2009.

Blium, Arlen Viktorovich, and Donna M. Farina. "Forbidden Topics: Early Soviet Censorship Directives." *Book History* 1 (1998): 268–82.

Bloom, Allan. *The Closing of the American Mind*. New York: Simon & Schuster, 1987.

———. "Responses to Fukuyama." *National Interest*, no. 16 (Summer 1989): 19–35.

Brand, Stewart. "We Owe It All to the Hippies." *Time*, March 1, 1995.

Breidlid, Anders, Fredrik Chr. Brøgger, Oyvind T. Gulliksen, and Torbjorn Sirevag. *American Culture: An Anthology*. 2nd ed. New York: Routledge, 2008.

Bremmer, Ian, and Mustafa Suleyman. "The AI Power Paradox: Can States Learn to Govern Artificial Intelligence—Before It's Too Late?" *Foreign Affairs*, August 16, 2023.

Brennan, Timothy. *Places of Mind: A Life of Edward Said*. New York: Farrar, Straus and Giroux, 2021.

Brill, Steven. *The Death of Truth: How Social Media and the Internet Gave Snake Oil Salesmen and Demagogues the Weapons They Needed to Destroy Trust and Polarize the World—and What We Can Do*. New York: Knopf, 2024.

———. "Trump, Palantir, and the Battle to Clean Up a Huge Army Procurement Swamp." *Fortune*, March 27, 2017.

Brock, Claire. "The Public Worth of Mary Somerville." *British Journal for the History of Science* 39, no. 2 (June 2006): 255–72.

Brodie, Janet Farrell. *The First Atomic Bomb: The Trinity Site in New Mexico*. Lincoln, Neb.: University of Nebraska Press, 2023.

Brooks, David. "The Sins of the Educated Class." *New York Times*, June 6, 2024.

Brown, Susan. "Soldier's Dad: 'He'll Always Be My Hero.'" *Times of Northwest Indiana*, October 2, 2011.

Bruckner, Pascal. *The Tears of the White Man: Compassion as Contempt*. Translated by William R. Beer. New York: Free Press, 1986.

Brudnick, Ida A. "Congressional Salaries and Allowances: In Brief." Congressional Research Service, June 27, 2024.

Brustein, Joshua. "Microsoft Wins $480 Million Army Battlefield Contract." *Bloomberg*, November 28, 2018.

Bryant, F. Carlene. *We're All Kin: A Cultural Study of a Mountain Neighborhood*. Knoxville, Tenn.: University of Tennessee Press, 1981.

Bubeck, Sébastien, et al. "Sparks of Artificial General Intelligence: Early Experiments with GPT-4." arXiv, March 22, 2023.

Buckley, Chris, and Didi Kirsten Tatlow. "Cultural Revolution Shaped Xi Jinping, from Schoolboy to Survivor." *New York Times*, September 24, 2015.

Burke, Kenneth. *Permanence and Change*. University of California Press, 1984.

Burnham, David. "The Computer, the Consumer, and Privacy." *New York Times*, March 4, 1984.

Burns, Ken, Lynn Novick, and Sarah Botstein. *The U.S. and the Holocaust*. PBS, September 18, 2022.

Bush, Vannevar. "As We May Think." *Atlantic Monthly*, July 1945.

———. *Modern Arms and Free Men*. New York: Simon & Schuster, 1949.

———. *Science: The Endless Frontier—A Report to the President*. Washington, D.C.: United States Government Printing Office, 1945.

Butler, Jack. "Does the Left Really Want to Argue That Enjoying *Lord of the Rings* Is 'Far-Right'?" *National Review*, July 19, 2024.

Byrne, John. "Northwest Indiana Medic Killed in Afghanistan." *Chicago Tribune*, October 1, 2011.

Caddell, Patrick. Interviewed in *American Experience: The Presidents*. PBS, November 11, 2002.

Carlyle, Thomas. *On Heroes: Hero-Worship, and the Heroic in History*. London: James Fraser, 1841.

Carter, Stephen L. *The Culture of Disbelief*. New York: Basic Books, 1993.

Cassy, John. "eToys Files for Bankruptcy." *Guardian*, February 28, 2001.

Ceraso, John, Irvin Rock, and Howard Gruber. "On Solomon Asch." In *The Legacy of Solomon Asch*. Hillsdale, N.J.: Lawrence Erlbaum Associates, 1990.

Channick, Robert. "Groupon Issues 'Going Concern' Warning as Chicago-Based Online Marketplace Terminates River North HQ Lease." *Chicago Tribune*, May 13, 2023.

Chapman, Lizette. "Inside Palantir's War with the U.S. Army." *Bloomberg*, October 28, 2016.

Chauncey, Sam, Jr. Letter to the Editor. *Yale Daily News*, November 29, 2017.

Cheyette, Fredric L. "Beyond Western Civilization: Rebuilding the Survey." *History Teacher* 10, no. 4 (August 1977): 535–38.

———. *Ermengard of Narbonne and the World of the Troubadours*. Ithaca, N.Y.: Cornell University Press, 2004.

Chicago Tribune. "Bermingham, Once Chicago Banker, Dies." July 14, 1958.

Child, Ben. "Mark Zuckerberg Rejects His Portrayal in The Social Network." *Guardian*, October 20, 2010.

China Daily. "President Xi's Speech on China-US Ties." September 24, 2015.

Chomsky, Noam, Ian Roberts, and Jeffrey Watumull. "The False Promise of ChatGPT." *New York Times*, March 8, 2023.

Christian, Brian. "How a Google Employee Fell for the Eliza Effect." *Atlantic*, June 21, 2022.

Clark, Kenneth. *Civilisation*. New York: Harper & Row, 1969.

———. "Remarks Announcing the Report of the National Performance Review and an Exchange with Reporters." Washington, D.C., September 7, 1993.

———. "Remarks on Signing the Federal Acquisition Streamlining Act of 1994." Washington, D.C., October 13, 1994.

———. State of the Union Address. Washington, D.C., January 23, 1996.

Coffey, John W. "State of Higher Education: Chaos." *Stanford Daily*, November 29, 1971.

Commentary. "The Study of Man: On Talcott Parsons." 1962.

Condliffe, Jamie. "Amazon Is Latest Tech Giant to Face Staff Backlash Over Government Work." *New York Times*, June 22, 2018.

Conger, Kate. "Google Plans Not to Renew Its Contract for Project Maven, a Controversial Pentagon Drone AI Imaging Program." *Gizmodo*, June 1, 2018.

Corcoran, Elizabeth. "Squaring Off in a Game of Checkers." *Washington Post*, August 14, 1994.

Cristancho, Sayra, and Graham Thompson. "Building Resilient Healthcare Teams: Insights from Analogy to the Social Biology of Ants, Honey Bees, and Other Social Insects." *Perspectives on Medical Education* 12, no. 1 (2023).

Curie, Eve. *Madame Curie*. Translated by Vincent Sheean. Garden City, N.Y.: Doubleday, Doran, 1937.

Daub, Adrian. "The Disruption Con: Why Big Tech's Favourite Buzzword Is Nonsense." *Guardian*, September 24, 2020.

Demick, Barbara, and David Pierson. "China Political Star Xi Jinping a Study in Contrasts." *Los Angeles Times*, February 11, 2012.

Denby, David. "In Darwin's Wake." *New Yorker*, July 21, 1997.

Dewey, John. "Pragmatic America." In *America's Public Philosopher: Essays on Social Justice, Economics, Education, and the Future of Democracy*, edited by Eric Thomas Weber. New York: Columbia University Press, 2021.

Die Welt. "Dieter Hildebrandt soll in NSDAP gewesen sein." June 30, 2007.

Doss, Erika. *Benton, Pollock, and the Politics of Modernism*. Chicago: University of Chicago Press, 1995.

Doward, Jamie. "A Gift-Horse in the Mouse." *Guardian*, October 23, 1999.

Dowd, Maureen. "The Ivy League Flunks Out." *New York Times*, December 9, 2023.

Draper, Robert. "Boondoggle Goes Boom." *New Republic*, June 18, 2013.

Drucker, Peter F. "The Coming of the New Organization." *Harvard Business Review*, January 1988.

Dudeck, Theresa Robbins. *Keith Johnstone: A Critical Biography*. London: Bloomsbury, 2013.

Duffy, Michael. "Hyman George Rickover, 1900–1986: They Broke the Mold." *Time*, July 21, 1986.

Dugatkin, Lee Alan. "Buffon, Jefferson, and the Theory of New World Degeneracy." *Evolution: Education and Outreach* 12 (2019).

Dunbar, Robin. "Co-Evolution of Neocortex Size, Group Size, and Language in Humans." *Behavioral and Brain Sciences* 16, no. 4 (1993).

Eagleton Institute of Politics. *Scientists in State Politics*. New Brunswick, N.J.: Rutgers University, 2023. eagleton.rutgers.edu/scientists-in-state-politics.

Economist. "Europe Faces a Painful Adjustment to Higher Defence Spending." February 22, 2024.

Einstein, Albert. Letter from Albert Einstein to Franklin D. Roosevelt, Peconic, N.Y., August 2, 1939. Franklin D. Roosevelt Presidential Library and Museum, Hyde Park, N.Y.

Eisenhower, Dwight D. Letter from Dwight D. Eisenhower to Edward J. Bermingham, February 28, 1951. Dwight D. Eisenhower Presidential Library, Abilene, Kans.

Emerson, Ralph Waldo. "Self-Reliance." In *Nature and Selected Essays*, edited by Larzer Ziff. Penguin Books, 2003.

Eschner, Kat. "Computers Are Great at Chess, But That Doesn't Mean the Game Is 'Solved.'" *Smithsonian Magazine*, February 10, 2017.

Eshel, Amir. *Jewish Memories, German Futures: Recent Debates in Germany About the Past*. Bloomington, Ind.: Indiana University, 2001.

Fahlenbrach, Rüdiger. "Founder-CEOs, Investment Decisions, and Stock Market Performance." *Journal of Financial and Quantitative Analysis* 44, no. 2 (April 2009): 439–66.

Fairhall, James. "The Case for the $435 Hammer." *Washington Monthly*, January 1, 1987.

Fandos, Nicholas. "In an Online World, a New Generation of Protesters Chooses Anonymity." *New York Times*, May 2, 2024.

Fano, Robert M. "Joseph Carl Robnett Licklider." In *Biographical Memoirs*, Vol. 3. Washington, D.C.: National Academies Press, 1998.

Farooq, Umar. "Pakistani Fertilizer Grows Both Taliban Bombs and Afghan Crops." *Christian Science Monitor*, May 9, 2013.

Ferguson, Niall. *Civilization: The West and the Rest*. New York: Penguin Books, 2011.

Filipovic, Jill. "I Was Wrong About Trigger Warnings." *Atlantic*, August 9, 2023.

Finney, John W. "Rickover, Father of Nuclear Navy, Dies at 86." *New York Times*, July 9, 1986.

Frank, Richard B. *Tower of Skulls: A History of the Asia–Pacific War: July 1937–May 1942*. New York: W. W. Norton, 2020.

Franke-Ruta, Garance. "Paul Harvey's 1978 'So God Made a Farmer' Speech." *Atlantic*, February 3, 2013.

Frankfurter Allgemeine. "Geistige Brandstiftung. Bubis wendet sich gegen Walser." October 13, 1998.

Freedman, Danny. "They're Getting Rid of 'Red Tape' in Washington. Literally." *Washington Post,* January 16, 2023.

Freire, Paulo. *Pedagogy of the Oppressed.* Translated by Myra Bergman Ramos. Penguin Books, 2017.

Freud, Sigmund. "Obsessive Actions and Religious Practices." In *Readings in Ritual Studies,* edited by Ronald L. Grimes. Upper Saddle River, N.J.: Prentice Hall, 1996.

Frisch, Karl von. *The Dance Language and Orientation of Bees.* Translated by Leigh E. Chadwick. Cambridge, Mass.: Belknap Press of Harvard University Press, 1967.

Fry, Jason. "eToys Story." *Wall Street Journal,* July 12, 1999.

Frye, Northrop. "The Decline of the West by Oswald Spengler." *Daedalus* 103, no. 1 (Winter 1974): 1–13.

Fukuyama, Francis. "The End of History?" *National Interest,* no. 16 (Summer 1989).
———. "Waltzing with (Leo) Strauss." *American Interest* 10, no. 4 (February 2015).

Future of Life Institute. "Pause Giant AI Experiments: An Open Letter." March 22, 2023. futureoflife.org/open-letter/pause-giant-ai-experiments.

Gaddis, John Lewis. *The Long Peace: Inquiries into the History of the Cold War.* New York: Oxford University Press, 1987.
———. *On Grand Strategy.* Penguin Books, 2019.

Gallup. "Confidence in Institutions." news.gallup.com/poll/1597/confidence-institutions.aspx.

Galton, Francis. "Vox Populi." *Nature* 75, no. 1949 (March 1907): 450–51.

Gayford, Martin. *Man with a Blue Scarf: On Sitting for a Portrait by Lucian Freud.* London: Thames & Hudson, 2019.

Girard, René. "Generative Scapegoating." In *Violent Origins: Walter Burket, René Girard, and Jonathan Z. Smith on Ritual Killing and Cultural Formation,* edited by Robert G. Hammerton-Kelly. Stanford, Calif.: Stanford University Press, 1987.

Gladwell, Malcolm. "The Tweaker." *New Yorker,* November 6, 2011.

Goethe, Johann Wolfgang von. *Faust.* 1808. Translated by Abraham Hayward and A. Bucheim. London: George Bell and Sons, 1892.

Goffman, Erving. *Asylums: Essays on the Social Situation of Mental Patients and Other Inmates.* London: Taylor & Francis, 2017.

Goh Keng Swee. *Report on the Ministry of Education, Singapore.* February 10, 1979.

Goldberg, Jeffrey. "The Obama Doctrine." *Atlantic*, April 2016.

Goldfarb, Brent, and David A. Kirsch. *Bubbles and Crashes: The Boom and Bust of Technological Innovation*. Stanford, Calif.: Stanford University Press, 2019.

Goldstein, Tom. "Neier Is Quitting Post at A.C.L.U.; He Denies Link to Defense of Nazis." *New York Times*, April 18, 1978.

Good, Irving John. "Speculations Concerning the First Ultraintelligent Machine." In *Advances in Computers*, Vol. 6, edited by Franz L. Alt and Morris Rubinoff. New York: Academic Press, 1965.

Gopinathan, Saravanan. "Singapore's Language Policies: Strategies for a Plural Society." *Southeast Asian Affairs* (1979): 280–95.

Gordon, Robert J. "The End of Economic Growth." *Prospect*, January 21, 2016.

Gore, Al. *Common Sense Government: Works Better and Costs Less*. National Performance Review (1998).

———. "Remarks at the National Performance Review Press Conference." October 26, 1993, Washington, D.C., C-SPAN.

Grady, Christine. "Institutional Review Boards: Purpose and Challenges." *CHEST* 148, no. 5 (November 2015): 1148–55.

Graeber, David. "The New Anarchists." *New Left Review* 13 (January/February 2002).

———. "Of Flying Cars and the Declining Rate of Profit." *Baffler*, no. 19 (March 2012).

Gray, Peter. "The Decline of Play and the Rise of Psychopathology in Children and Adolescents." *American Journal of Play* 3, no. 4 (2011).

Green, Heather. "The Great Yuletide Shakeout." *Businessweek*, November 1, 1999.

Greene, Jay. "Amazon Bans Police Use of Its Facial-Recognition Technology for a Year." *Washington Post*, June 10, 2020.

Greenspan, Alan. "Remarks by Chairman Alan Greenspan: At the Annual Dinner and Francis Boyer Lecture of the American Enterprise Institute for Public Policy Research." Washington, D.C., December 5, 1996.

Gregory I. *The Life of Our Most Holy Father S. Benedict*. Rome: 1895.

Guerrieri, Matthew. *The First Four Notes: Beethoven's Fifth and the Human Imagination*. New York: Vintage Books, 2014.

Gutmann, Amy. "Democratic Citizenship." *Boston Review*, October 1, 1994.

Habermas, Jürgen. *Legitimation Crisis*. Translated by Thomas McCarthy. Cambridge, U.K.: Polity Press, 1976.

Hall, Andrew B. *Who Wants to Run? How the Devaluing of Political Office Drives Polarization*. Chicago: University of Chicago Press, 2019.

Handler, Edward. "'Nature Itself Is All Arcanum': The Scientific Outlook of John

Adams." *Proceedings of the American Philosophical Society* 120, no. 3 (June 1976): 216–29.

Hanna, Graeme. "'Stop Working with Pentagon'—OpenAI Staff Face Protests." *ReadWrite*, February 13, 2024.

Harari, Yuval Noah. *Sapiens: A Brief History of Humankind*. New York: HarperCollins, 2015.

Harris, Gareth. "Mary Beard BBC Segment on Kenneth Clark's Civilisation Renews Debate About Its Eurocentricity." *Art Newspaper*, April 29, 2024.

Harris, Robin. *Not for Turning: The Life of Margaret Thatcher*. New York: Thomas Dunne Books, 2013.

Harris, Shane. "Palantir Wins Competition to Build Army Intelligence System." *Washington Post*, March 26, 2019.

Hartocollis, Anemona, Stephanie Saul, and Vimal Patel. "At Harvard, a Battle Over What Should Be Said About the Hamas Attacks." *New York Times*, October 10, 2023.

Harwell, Drew. "Amazon Extends Ban on Police Use of Its Facial Recognition Technology Indefinitely." *Washington Post*, May 18, 2021.

Hay, Denys. *Europe: The Emergence of an Idea*. 2nd ed. Edinburgh: Edinburgh University Press, 1968.

Heinrich, Thomas. "Cold War Armory: Military Contracting in Silicon Valley." *Enterprise and Society* 3, no. 2 (June 2002): 247–84.

Heller, Ágnes. *Beyond Justice*. Oxford: Blackwell, 1987.

Helman, Christopher, and Hank Tucker. "The War in Afghanistan Cost America $300 Million per Day for 20 Years, with Big Bills Yet to Come." *Forbes*, August 16, 2021.

Henderson, Rob. "'Luxury Beliefs' Are the Latest Status Symbol for Rich Americans." *New York Post*, August 17, 2019.

———. *Troubled: A Memoir of Foster Care, Family, and Social Class*. New York: Gallery Books, 2024.

Henshall, Will. "The U.S. Military's Investments into Artificial Intelligence Are Skyrocketing," *Time*, March 29, 2024.

Herman, Arthur. *Freedom's Forge: How American Business Produced Victory in World War II*. New York: Random House, 2013.

———. "What if Apple Designed an iFighter?" *Wall Street Journal*, July 23, 2012.

Hersey, John. *Hiroshima*. New York: Vintage Books, 2020.

Hewlett, Richard G., and Francis Duncan. *Nuclear Navy, 1946–1962*. Chicago: University of Chicago Press, 1974.

Hirsch, E. D., Jr. *Cultural Literacy: What Every American Needs to Know*. New York: Vintage, 1988.

Hirsch, Lauren. "One Law Firm Prepared Both Penn and Harvard for Hearing on Antisemitism." *New York Times*, December 8, 2023.

Hoffheimer, Michael H. *Justice Holmes and The Natural Law*. New York: Routledge, 2013.

Hofstadter, Douglas. "Gödel, Escher, Bach, and AI." *Atlantic*, July 8, 2023.

Holusha, John. "Taiichi Ohno, Whose Car System Aided Toyota's Climb, Dies at 78." *New York Times*, May 31, 1990.

Homer. *The Odyssey of Homer*. Translated by George Herbert Palmer. Cambridge, Mass.: Houghton Mifflin, 1949.

Hoover, Herbert. *The Memoirs of Herbert Hoover: Years of Adventure, 1874–1920*. New York: Macmillan, 1953.

Huet, Ellen. "Protesters Gather Outside OpenAI Headquarters." *Bloomberg*, February 13, 2024.

Huntington, Samuel P. "The Clash of Civilizations?" *Foreign Affairs* 72, no. 3 (Summer 1993): 22–49.

———. "Dead Souls: The Denationalization of the American Elite." *National Interest* (Spring 2004).

Illmer, Andreas. "German Writer Martin Walser Dies Aged 96." *Deutsche Welle*, July 28, 2023.

Impelli, Matthew. "Jerome Powell Salary Admission Sparks Debate." *Newsweek*, February 7, 2023.

Irish Times. "Le Pen Scores Own Goal with Team Slur." June 25, 1996.

Isaacson, Walter. *Benjamin Franklin: An American Life*. New York: Simon & Schuster, 2003.

———. *The Innovators: How a Group of Hackers, Geniuses, and Geeks Created the Digital Revolution*. New York: Simon & Schuster, 2015.

———. *Steve Jobs*. New York: Simon & Schuster, 2013.

Ismay, John. "'We Hated What We Were Doing': Veterans Recall Firebombing Japan." *New York Times*, March 9, 2020.

Issa, Darrel and Jason Chaffetz. Letter from Darrel Issa and Jason Chaffetz to Leon E. Panetta, August 1, 2012.

Jacobsen, Annie. *Operation Paperclip: The Secret Intelligence Program That Brought Nazi Scientists to America*. New York: Little, Brown, 2014.

———. "Palantir's God's-Eye View of Afghanistan." *Wired*, January 20, 2021.

Jefferson, Thomas. Letter from Thomas Jefferson to Harry Innes, Philadelphia, March 7, 1791. In *The Papers of Thomas Jefferson*, Vol. 19. Princeton, N.J.: Princeton University Press, 1974.

Jégo, Yves. "Emmanuel Macron et le reniement de la culture française." *Le Figaro*, February 6, 2017.

Jensen, Jeff. "Toby Lenk." *Advertising Age*, June 1, 1998.

Johnson, Ian. "In Singapore, Chinese Dialects Revive After Decades of Restrictions." *New York Times*, August 26, 2017.

Johnstone, Keith. *Impro: Improvisation and the Theatre*. New York: Routledge, 1981.

Junge, Traudl. *Until the Final Hour: Hitler's Last Secretary*, edited by Melissa Müller and translated by Anthea Bell. New York: Arcade, 2004.

Kabaservice, Geoffrey. *The Guardians: Kingman Brewster, His Circle, and the Rise of the Liberal Establishment*. New York: Henry Holt, 2004.

Kalven, Harry, Jr. *The Negro and the First Amendment*. Columbus: Ohio State University Press, 1965.

Kamenetzky, David A. "The Debate on National Identity and the Martin Walser Speech: How Does Germany Reckon with Its Past?" *SAIS Review* 19, no. 2 (Summer–Fall 1999): 257–66.

Karmel, Pepe, ed. *Jackson Pollock: Interviews, Articles, and Reviews*. New York: Museum of Modern Art, 1999.

Karp, Alexander C. "Our Oppenheimer Moment: The Creation of A.I. Weapons." *New York Times*, July 25, 2023.

Karp, Alexander C., and Nicholas W. Zamiska. "New Weapons Will Eclipse Atomic Bombs. Their Builders Should Ask Themselves This Question." *Washington Post*, June 25, 2024.

———. "Silicon Valley Has a Harvard Problem." *Time*, February 12, 2024.

Kasparov, Garry. *Deep Thinking: Where Machine Intelligence Ends and Human Creativity Begins*. New York: PublicAffairs, 2017.

Keeley, Lawrence H. *War Before Civilization: The Myth of the Peaceful Savage*. New York: Oxford University Press, 1996.

Kelleher, Thomas J., Jr., et al. *Smith, Currie, and Hancock's Federal Government Construction Contracts*. Hoboken, N.J.: Wiley, 2010.

Kennedy, Paul. *The Rise and Fall of the Great Powers: Economic Change and Military Conflict from 1500 to 2000*. New York: Random House, 1989.

Kerouac, Jack. *On the Road*. New York: Penguin Books, 1976.

Kimball, Bruce, ed. *The Liberal Arts Tradition: A Documentary History*. Lanham, Md.: University Press of America, 2010.

Kimball, Roger. "Institutionalizing Our Demise: America vs. Multiculturalism." *New Criterion*, June 2004.

———. "The Perils of Designer Tribalism." *New Criterion*, April 2001.

King, Charles. *Gods of the Upper Air: How a Circle of Renegade Anthropologists Reinvented Race, Sex, and Gender in the Twentieth Century*. New York: Anchor Books, 2020.

Kintner, E. E. "Admiral Rickover's Gamble." *Atlantic*, January 1959.

Kissinger, Henry. Foreword to *From Third World to First: The Singapore Story: 1965–2000*, by Lee Kuan Yew. New York: HarperCollins, 2000.

Kitchen, Lynn W., David W. Vaughn, and Donald R. Skillman. "Role of U.S. Military Research Programs in the Development of U.S. Food and Drug Administration–Approved Antimalarial Drugs." *Clinical Infectious Diseases* 43, no. 1 (2006): 67–71.

Klebnikov, Sergei. "U.S. Tech Stocks Are Now Worth More Than $9 Trillion, Eclipsing the Entire European Stock Market." *Forbes*, December 15, 2020.

Kovach, Thomas, and Martin Walser. *The Burden of the Past: Martin Walser on Modern German Identity: Texts, Contexts, Commentary*. Rochester, N.Y.: Camden House, 2008.

Krcmaric, Daniel, Stephen C. Nelson, and Andrew Roberts. "Billionaire Politicians: A Global Perspective." *Perspectives on Politics*, October 25, 2023.

Kris, Ernst. *Psychoanalytic Explorations in Art*. New York: International Universities' Press, 1952.

Kristol, Irving. "Countercultures." *Commentary*, December 1994.

Kuralt, Charles. "Edward R. Murrow." *The North Carolina Historical Review* 48, no. 2 (April 1971): 161–170.

LaFrance, Adrienne. "The Rise of Techno-Authoritarianism." *Atlantic*, January 30, 2024.

Lassiter, Matthew D. "Who Speaks for the Silent Majority?" *New York Times*, November 2, 2011.

Lee, Joon Mahn, Jongsoo Kim, and Joonhyung Bae. "Founder CEOs and Innovation: Evidence from S&P 500 Firms." SSRN, February 17, 2016.

Lee Kuan Yew. "In His Own Words: Higher Pay Will Attract Most Talented Team, So Country Can Prosper." *Straits Times*, November 1, 1994.

———. *Lee Kuan Yew: The Grand Master's Insights on China, the United States, and the World*, edited by Graham Allison and Robert D. Blackwill. Cambridge, Mass.: MIT Press, 2020.

———. "Speech at the National Day Rally," Kallang Theatre, Singapore, August 17, 1986, Singapore National Archives.

———. "Speech at the Opening of the 'Speak Mandarin Campaign.'" Singapore Conference Hall, Singapore, September 21, 1984, Singapore National Archives.

———. "Speech at the 28th Anniversary of Liquor Retailers' Association." Chinese Chamber of Commerce, Singapore, October 3, 1965, Singapore National Archives.

Leith, Suzette. "Civ: Enlightenment ... or the Black Death?" *Stanford Daily*, May 17, 1966.

Lemoine, Blake. "Explaining Google." Medium, May 30, 2019.

Lepore, Jill. "The X-Man." *New Yorker*, September 11, 2023.

Leslie, Stuart W. "The Biggest 'Angel' of Them All: The Military and the Making of Silicon Valley." In *Understanding Silicon Valley: The Anatomy of an Entrepreneurial Region*, edited by Martin Kenney. Stanford, Calif.: Stanford University Press, 2000.

Levinson, Robert. "The F-35's Global Supply Chain." *Businessweek*, September 1, 2011.

Lévi-Strauss, Claude. *Tristes Tropiques*. Translated by John Weightman and Doreen Weightman. New York: Penguin Books, 2012.

Levy, Steven. *Hackers: Heroes of the Computer Revolution*. Sebastopol, Calif.: O'Reilly, 2010.

Licklider, J. C. R. "Man-Computer Symbiosis." *IRE Transactions on Human Factors in Electronics*, no. 1 (March 1960): 4–11.

Lindauer, Martin. "House-Hunting by Honey Bee Swarms." Translated by P. Kirk Visscher, Karin Behrens, and Susanne Kuehnholz. *Journal of Comparative Physiology* 37 (1955).

Link, Perry. "China: The Anaconda in the Chandelier." *New York Review of Books*, April 11, 2002.

Linklater, Richard, dir. *Before Sunset*. 2004; Burbank, Calif.: Warner Independent Pictures.

Long, Heather. "Who Is Jerome Powell, Trump's Pick for the Nation's Most Powerful Economic Position?" *Washington Post*, November 2, 2017.

Lorenz, Konrad Z. *King Solomon's Ring*. New York: Thomas Y. Crowell, 2020.

Losey, Stephen. "F-35s to Cost $2 Trillion as Pentagon Plans Longer Use, Says Watchdog." *Defense News*, April 15, 2024.

Ludendorff, Erich. *The "Total" War*, edited by Herbert Lawrence. London: Friends of Europe, 1936.

Lundberg, Ferdinand. *America's 60 Families*. New York: Vanguard Press, 1937.

MacIntyre, Alasdair. *After Virtue: A Study in Moral Theory*. 3rd ed. South Bend, Ind.: University of Notre Dame Press, 2007.

Madison, James. Letter from James Madison to Thomas Jefferson, June 19, 1786. In *The Writings of James Madison*, Vol. 2, edited by Gaillard Hunt. New York: G. P. Putnam's Sons, 1901.

———. *The Writings of James Madison*, Vol. 3, edited by Gaillard Hunt. New York: G. P. Putnam's Sons, 1902.

Mallapaty, Smriti, Jeff Tollefson, and Carissa Wong. "Do Scientists Make Good Presidents?" *Nature*, June 6, 2024.

Martinez, Hannah J. "The Graduating Class of 2023 by the Numbers." *Harvard Crimson*, 2023.

Martinson, Jane, and Larry Elliott. "The Year Dot.com Turned into Dot.bomb." *Guardian*, December 29, 2000.

Maxwell, William J., ed. *James Baldwin: The FBI File*. New York: Arcade, 2017.

Mazzucato, Mariana. *The Entrepreneurial State: Debunking Public vs. Private Sector Myths*. London: Anthem Press, 2013.

McCarthy, J., M. L. Minsky, N. Rochester, and C. E. Shannon. "A Proposal for the Dartmouth Summer Research Project on Artificial Intelligence." August 31, 1955 (reproduced in *AI Magazine* 27, no. 4 (Winter 2006): 12–14).

McCullough, Brian. *How the Internet Happened: From Netscape to the iPhone*. New York: Liveright, 2018.

McLaughlin, Andrew C., et al. *The Study of History in Schools: Report to the American Historical Association*. New York: Macmillan, 1899.

McNeill, William H. "Western Civ in World Politics: What We Mean by the West." *Orbis* 41, no. 4 (Fall 1997): 513–24.

Meacham, Jon. *Thomas Jefferson: The Art of Power*. New York: Random House, 2013.

Mencken, H. L. "Ludendorff." *Atlantic*, June 1917.

The Metropolitan Museum of Art. *Monet's Years at Giverny: Beyond Impressionism*. New York: Harry N. Abrams Publishers, 1978.

Meyer, Edith Patterson. *Dynamite and Peace: The Story of Alfred Nobel*. Boston: Little, Brown, 1958.

Milgram, Stanley. *Obedience to Authority: An Experimental View*. New York: Harper Perennial, 2009.

Mills, D. Quinn. "Who's to Blame for the Bubble?" *Harvard Business Review*, May 2001.

Mirowski, Piotr, Juliette Love, Kory Mathewson, and Shakir Mohamed. "A Robot Walks into a Bar: Can Language Models Serve as Creativity Support Tools for Comedy? An Evaluation of LLMs' Humour Alignment with Comedians." arXiv, June 3, 2024.

Mishra, Pankaj. "The Reorientations of Edward Said." *New Yorker*, April 19, 2021.

Mothershed, Airon A. "The $435 Hammer and $600 Toilet Seat Scandals: Does Media Coverage of Procurement Scandals Lead to Procurement Reform?" *Public Contract Law* 41, no. 4 (Summer 2012): 855–80.

Mouat, Jeremy, and Ian Phimister. "The Engineering of Herbert Hoover." *Pacific Historical Review* 77, no. 4 (November 2008): 553–84.

Mozur, Paul. "Inside China's Dystopian Dreams: A.I., Shame, and Lots of Cameras." *New York Times*, July 8, 2018.

Mozur, Paul, and Adam Satariano. "A.I. Begins Ushering in an Age of Killer Robots." *New York Times*, July 2, 2024.

Muehlhauser, Luke. "What Should We Learn from Past AI Forecasts?" Open Philanthropy, May 2016. openphilanthropy.org/research/what-should-we-learn-from-past-ai-forecasts.

Murray, Pauli. *Song in a Weary Throat: Memoir of an American Pilgrimage*. New York: Liveright, 2018.

Mydans, Seth. "Singapore Announces 60 Percent Pay Raise for Ministers." *New York Times*, April 9, 2007.

Nakashima, Ellen, and Reed Albergotti. "The FBI Wanted to Unlock the San Bernardino Shooter's iPhone. It Turned to a Little-Known Australian Firm." *Washington Post*, April 14, 2021.

National Defense Industrial Association (NDIA). "Army 'Rapid Equipping Force' Taking Root, Chief Says." *National Defense*, October 1, 2006.

National Physical Laboratory. "Tracking People by Their 'Gait Signature.'" September 20, 2012.

National Student Clearinghouse. "Computer Science Has Highest Increase in Bachelor's Earners." National Student Clearinghouse, May 27, 2024.

Negroponte, Nicholas. "Big Idea Famine." *Journal of Design and Science*, no. 3 (February 2018).

Neier, Aryeh. *Defending My Enemy: American Nazis, the Skokie Case, and the Risks of Freedom*. New York: E. P. Dutton, 1979.

Nevers, Kevin. "'He Didn't Hesitate': Airborne Medic Jim Butz Dies a Hero in Afghanistan." *Chesterton* (Ind.) *Tribune*, October 3, 2011.

Newman, John. "Singapore's *Speak Mandarin Campaign*: The Educational Argument." *Southeast Asian Journal of Social Science* 14, no. 2 (1986): 52–67.

New York Times. "Admiral Rickover and the Trinkets." May 24, 1985.

———. "The Dot-Com Bubble Bursts." December 24, 2000.

———. "Rickover Tells Lehman He Gave Away Gifts." June 11, 1985.

Ng, Patrick Chin Leong. "A Study of Attitudes Towards the Speak Mandarin Campaign in Singapore." *Intercultural Communication Studies* 23, no. 3 (2014): 53–65.

Nixon, Richard. "Address to the Nation on the War in Vietnam." November 3, 1969. Washington, D.C.

Noonan, Peggy. "How Trump Lost Half of Washington." *Wall Street Journal*, April 25, 2019.

———. "A Six-Month AI Pause? No, Longer Is Needed." *Wall Street Journal*, March 30, 2023.

———. "The Uglyfication of Everything." *Wall Street Journal*, May 2, 2024.

Nussbaum, Martha. "Patriotism and Cosmopolitanism." *Boston Review*, October 1, 1994.

Nye, Joseph S., Jr. *Soft Power: The Means to Success in World Politics*. New York: PublicAffairs, 2004.

Office of the Under Secretary of Defense. *U.S. Department of Defense Fiscal Year 2024 Budget Request*. March 2023. comptroller.defense.gov/Budget-Materials/Budget2024.

Ohno, Taiichi. *Toyota Production System: Beyond Large-Scale Production*. Boca Raton, Fla.: CRC Press, 1988.

Olusoga, David. "Civilisation Revisited." *Guardian*, February 3, 2018.

Oppenheimer, J. Robert. "Physics in the Contemporary World." *Bulletin of the Atomic Scientists* 4, no. 3 (1948): 65–86.

Orton, Brad. "National Performance Review." October 26, 1993, Old Executive Office Building, Washington, D.C., C-SPAN.

Orwell, George. *1984*. New York: Penguin, 2023.

Osnos, Evan. "What Did China's First Daughter Find in America?" *New Yorker*, April 6, 2015.

———. "Xi Jinping's Historic Bid at the Communist Party Congress." *New Yorker*, October 23, 2022.

Packer, George. "No Death, No Taxes." *New Yorker*, November 20, 2011.

Packer, Herbert L., et al. *The Study of Education at Stanford: Report to the University*, Vol. 2. Stanford, Calif.: Stanford University, November 1968.

Pagé, Suzanne, Marianne Mathieu, and Angeline Scherf. *Monet—Mitchell*. New Haven, Conn.: Yale University Press, 2022.

Parisi, Giorgio. *In a Flight of Starlings: The Wonder of Complex Systems*. Translated by Simon Carnell. New York: Penguin Books, 2023.

Parsons, Talcott. "Certain Primary Sources and Patterns of Aggression in the Social Structure of the Western World." In *Essays in Sociological Theory*, Rev. ed. Glencoe, Ill.: Free Press, 1954.

Perlow, Leslie A., Constance Noonan Hadley, and Eunice Eun. "Stop the Meeting Madness." *Harvard Business Review*, July–August 2017.

Perlroth, Nicole. "The Groupon IPO: By the Numbers." *Forbes*, June 2, 2011.

Perlstein, Rick. *Nixonland: The Rise of a President and the Fracturing of America*. New York: Scribner, 2008.

Perry, Robert. *A History of Satellite Reconnaissance*. U.S. National Reconnaissance Office, October 1973.

Peters, Jay. "IBM Will No Longer Offer, Develop, or Research Facial Recognition Technology." *Verge*, June 8, 2020.

Pfeiffer, John. *The Thinking Machine*. Philadelphia: J. B. Lippincott, 1962.

Pinker, Steven. *The Better Angels of Our Nature: Why Violence Has Declined*. New York: Penguin Books, 2011.

Planck, Max. *Scientific Autobiography and Other Papers*. Translated by Frank Gaynor. New York: Philosophical Library, 1949.

Plant, Raymond. "Jürgen Habermas and the Idea of Legitimation Crisis." *European Journal of Political Research* 10 (1982): 341–52.

Plato. *The Republic*. Translated by Desmond Lee. New York: Penguin Books, 2007.

Polenberg, Richard, ed. *In the Matter of J. Robert Oppenheimer: The Security Clearance Hearing*. Ithaca, N.Y.: Cornell University Press, 2002.

Polmar, Norman, and Thomas B. Allen. *Rickover: Controversy and Genius: A Biography*. New York: Simon & Schuster, 1982.

Porter, Catherine. "Cheers, Fears, and 'Le Wokisme': How the World Sees U.S. Campus Protests." *New York Times*, May 3, 2024.

Powell, Jerome. "The Honorable Jerome H. Powell." Interview by David M. Rubenstein. Economic Club of Washington D.C., February 7, 2023.

Quinn, Roswell. "Rethinking Antibiotic Research and Development: World War II and the Penicillin Collaborative." *American Journal of Public Health* 103, no. 3 (2013): 426–34.

Raffoul, Amanda, et al. "Social Media Platforms Generate Billions of Dollars in Revenue from U.S. Youth: Findings from a Simulated Model." *PLoS ONE*, December 27, 2023.

Rainey, Clint. "P(doom) Is AI's Latest Apocalypse Metric. Here's How to Calculate Your Score." *Fast Company*, December 7, 2023.

Rajan, Raghuram, and Julie Wulf. "The Flattening Firm: Evidence from Panel Data on the Changing Nature of Corporate Hierarchies." Working Paper No. 9633. National Bureau of Economic Research, April 2003.

Ramzy, Austin. "Xi Jinping on 'House of Cards' and Hemingway." *New York Times*, September 23, 2015.

Rauchhaus, Robert. "Evaluating the Nuclear Peace Hypothesis." *Journal of Conflict Resolution* 53, no. 2 (April 2009): 258–77.

Rawls, John. *Political Liberalism*. New York: Columbia University Press, 1993.

———. "The Priority of Right and Ideas of the Good." *Philosophy and Public Affairs* 17, no. 4 (Autumn 1988): 253–76.

Remnick, David. "The Scholar of Comedy." *New Yorker*, April 28, 2024.

Renan, Ernest. *What Is a Nation? And Other Political Writings*. Translated and edited by M. F. N. Giglioli. New York: Columbia University Press, 2018.

Reynolds, Winston A. "The Burning Ships of Hernán Cortés." *Hispania* 42, no. 3 (September 1959): 317–24.

Rickover, Hyman G. Interview by Diane Sawyer. *60 Minutes*, CBS, 1984.

Rigden, John S. *Rabi: Scientist and Citizen*. New York: Basic Books, 1987.

Rigolot, François. "Curiosity, Contingency, and Cultural Diversity: Montaigne's Readings at the Vatican Library." *Renaissance Quarterly* 64, no. 3 (Fall 2011): 847–74.

Rogers, Alex. "The MRAP: Brilliant Buy, or Billions Wasted?" *Time*, October 2, 2012.

Rohlfs, Chris, and Ryan Sullivan. "Why the $600,000 Vehicles Aren't Worth the Money." *Foreign Affairs*, July 26, 2012.

Roose, Kevin. "Bing's A.I. Chat: 'I Want to Be Alive.'" *New York Times*, February 16, 2023.

Roosevelt, Theodore. "Citizenship in a Republic: Address Delivered at the Sorbonne, Paris, April 23, 1910." In *Presidential Addresses and State Papers and European Addresses: December 8, 1908, to June 7, 1910*. New York: Review of Reviews, 1910.

Rosen, Charles. "Review of Civilisation, by Kenneth Clark." *New York Review of Books*, May 7, 1970.

Rosenbaum, David E. "Remaking Government: Few Disagree with Clinton's Overall Goal, but History Shows the Obstacles Ahead." *New York Times*, September 8, 1993.

Rusli, Evelyn M. "Zynga's Value, at $7 Billion, Is Milestone for Social Gaming." *New York Times*, December 15, 2011.

Sabato, Larry J. *Feeding Frenzy: How Attack Journalism Has Transformed American Politics*. New York: Free Press, 1993.

Said, Edward. *Orientalism*. New York: Vintage Books, 1979.

Sallust. *The War with Catiline*. Translated by J. C. Rolfe and revised by John T. Ramsey. Loeb Classical Library. Cambridge, Mass.: Harvard University Press, 2013.

Salovey, Peter. "Free Speech, Personified." *New York Times*, November 26, 2017.

Sandberg-Diment, Erik. "Hardware Review: Apple Weighs in with Its Macintosh." *New York Times*, January 24, 1984.

Sandel, Michael. *Liberalism and the Limits of Justice*. 2nd ed. Cambridge, U.K.: Cambridge University Press, 1998.

———. *What Money Can't Buy: The Moral Limits of Markets*. New York: Farrar, Straus and Giroux, 2012.

Sapolsky, Harvey M., and Michael Schrage. "More Than Technology Needed To Defeat Roadside Bombs." *National Defense*, April 2012.

Savitz, Eric J. "Groupon Stock Craters. The Turnaround Is Taking Longer Than Hoped." *Barron's*, November 10, 2023.

Sax, Boria. *Imaginary Animals: The Monstrous, the Wondrous and the Human*. London: Reaktion Books, 2013.

Scarborough, Rowan. "Soldier Battling Bombs Irked by Software Switch." *Washington Times*, July 22, 2012.

Schell, Orville. "What Xi Jinping's Seattle Speech Might Mean for the U.S." *Foreign Policy*, September 23, 2015.

Schelling, Thomas. *Arms and Influence*. New Haven, Conn.: Yale University Press, 2020.

Schlosser, Eric. *Command and Control: Nuclear Weapons, the Damascus Accident, and the Illusion of Safety*. New York: Penguin Press, 2013.

Schödel, Kathrin. "Normalising Cultural Memory? The 'Walser-Bubis Debate' and Martin Walser's Novel *Ein springender Brunnen*." In *Recasting German Identity: Culture, Politics, and Literature in the Berlin Republic*, edited by Stuart Taberner and Frank Finlay. Rochester, N.Y.: Camden House, 2002.

Schulz, Kathryn. "The Many Lives of Pauli Murray." *New Yorker*, April 10, 2017.

Scruton, Roger. "Animal Rights." *City Journal* (Summer 2000).

Seeley, Thomas D. *Honeybee Democracy*. Princeton, N.J.: Princeton University Press, 2010.

———. "Martin Lindauer (1918–2008)." *Nature*, December 11, 2008.

Seeley, T.D., S. Kunholz, and R.H. Seeley. "An Early Chapter in Behavioral Physiology and Sociobiology: The Science of Martin Lindauer." *Journal of Comparative Physiology*, 188 (July 2002).

Sennett, Richard. "The Identity Myth." *New York Times*, January 30, 1994.

Shaban, Hamza. "Google Parent Alphabet Reports Soaring Ad Revenue, Despite YouTube Backlash." *Washington Post*, February 1, 2018.

Shane, Leo, III. "Why One Lawmaker Keeps Pushing for a New Military Draft." *Military Times*, March 30, 2015.

Shane, Scott, and Daisuke Wakabayashi. "'The Business of War': Google Employees Protest Work for the Pentagon." *New York Times*, April 4, 2018.

Shatz, Adam. "'Orientalism,' Then and Now." *New York Review of Books*, May 20, 2019.

Shell, Jason. "How the IED Won: Dispelling the Myth of Tactical Success and Innovation." *War on the Rocks*, May 1, 2017.

Sherwood, Harriet. "Hamas Says 250 People Held Hostage in Gaza." *Guardian*, October 16, 2023.

Shiller, Robert J. *Irrational Exuberance*. New York: Crown, 2006.

Sigel, Efrem. "Harvard, Yale Students to Issue New Invitations to Gov. Wallace." *Harvard Crimson*, September 25, 1963.

———. "New Wallace Invitation Expected at Yale Today." *Harvard Crimson*, September 24, 1963.

Silver, Nate. *On the Edge: The Art of Risking Everything*. New York: Penguin Press, 2024.

Simon, Herbert A. *The New Science of Management Decision*. New York: Harper & Brothers, 1960.

Simon, Herbert A., and Allen Newell. "Heuristic Problem Solving: The Next Advance in Operations Research." *Operations Research* 6, no. 1 (January–February 1958): 1–10.

Simonite, Tom. "Behind the Rise of China's Facial-Recognition Giants." *Wired*. September 3, 2019.

Sims, David. "No, Really, I'm Awful." *Atlantic*, April 26, 2023.

Singer, Peter. *Animal Liberation: A New Ethics for Our Treatment of Animals*. New York: New York Review Books, 1975.

Sledge, Matt, and Ramon Antonio Vargas. "Palantir's Crime-Fighting Software Causes Stir in New Orleans; NOPD Rebuts Civil Liberties Concerns." *Times-Picayune*, March 1, 2018.

Slomovic, Anna. *Anteing Up: The Government's Role in the Microelectronics Industry*. Santa Monica, Calif.: RAND Corporation, December 16, 1988.

Smith, Emily Esfahani. "The Friendship That Changed Art." *Artists Magazine* 35, no. 6 (July/August 2018).

Smith, James K. A. "Reconsidering 'Civil Religion.'" *Comment*, May 11, 2017.

Smith, Jean Edward. *FDR*. New York: Random House, 2008.

Sneed, Annie. "Computer Beats Go Champion for First Time." *Scientific American*, January 27, 2016.

Soapes, Thomas. Interview with Hans A. Bethe. Dwight D. Eisenhower Library, Abilene, Kans., November 3, 1977.

Sokolove, Michael. "How to Lose $850 Million—and Not Really Care." *New York Times Magazine*, June 9, 2002.

Solomon, Charles. "Two States—One Nation?" *Los Angeles Times*, November 17, 1991.

Somers, James. "The Man Who Would Teach Machines to Think." *Atlantic*, November 15, 2013.

Somerville, Mary. *On the Connexion of the Physical Sciences*. London: John Murray, 1834.

Spengler, Oswald. *The Decline of the West*. New York: Oxford University Press, 1991.

Spiegel. "Martin Walser Bereut Verhalten Gegenüber Ignatz Bubis." March 16, 2007.

Spitz, Lewis W. "Beyond Western Civilization: Rebuilding the Survey." *History Teacher* 10, no. 4 (August 1977): 515–24.

Stanley, Jay. "New Orleans Program Offers Lessons in Pitfalls of Predictive Policing." *American Civil Liberties Union*, 2018.

Starkie, Thomas. *A Practical Treatise on the Law of Evidence, and Digest of Proofs, in Civil and Criminal Proceedings*, Vol. 1. Boston: Wells & Lilly, 1826.

Stewart, Emilie B. "Survey of PRC Drone Swarm Inventions." China Aerospace Studies Institute, U.S. Air Force, October 2023.

Stout, David. "Solomon Asch Is Dead at 88; A Leading Social Psychologist." *New York Times*, February 29, 1996.

Strauss, Leo. *What Is Political Philosophy?* Chicago: University of Chicago Press, 1959.

Suleyman, Mustafa. *The Coming Wave: Technology, Power, and the Twenty-First Century's Greatest Dilemma*. With Michael Bhaskar. New York: Crown, 2023.

Sullivan, Walter. "65% in Test Blindly Obey Order to Inflict Pain." *New York Times*, October 26, 1963.

Summers, Lawrence. Interview by David Remnick. *New Yorker Radio Hour*. NPR, May 3, 2024.

Surowiecki, James. *The Wisdom of Crowds*. New York: Anchor Books, 2005.

Sutter, Daniel. "Media Scrutiny and the Quality of Public Officials." *Public Choice* 129 (2006): 25–40.

Swensen, David. "A Conversation with David Swensen." Interview by Robert E. Rubin. Council on Foreign Relations, November 14, 2017.

Tarnoff, Ben. "Tech Workers Versus the Pentagon." *Jacobin*, June 6, 2018.

Taylor, Paul. "How to Spend Europe's Defense Bonanza Intelligently." *Politico*, September 2, 2022.

Tetlock, Philip E. *Expert Political Judgment: How Good Is It? How Can We Know?* Princeton, N.J.: Princeton University Press, 2005.

Thomas, Dylan. *The Collected Poems of Dylan Thomas: Original Edition*. New York: New Directions, 2010.

Thomas, Sean. "Are We Ready for P(doom)?" *Spectator*, March 4, 2024.

Tiku, Nitasha. "The Google Engineer Who Thinks the Company's AI Has Come to Life." *Washington Post*, June 11, 2022.

———. "Google Fired Engineer Who Said Its AI Was Sentient." *Washington Post*, July 22, 2022.

Time. "Cosmoclast Einstein." July 1, 1946.

———. "Man and Woman of the Year: The Middle Americans." January 5, 1970.

———. "Patterns in Chaos." Review of *The Decline of the West: Perspectives of World History*, by Oswald Spengler. December 10, 1928.

———. "The Press: In a Corner, on the 13th Floor." July 22, 1946.

———. "Public Figures and Their Private Lives." August 22, 1969.

———. "The Roosevelt Week." February 5, 1934.

Treaster, Joseph B. "Brewster Doubts Fair Black Trials." *New York Times*, April 25, 1970.

Turchin, Peter. *End Times: Elites, Counter-Elites, and the Path of Political Disintegration*. New York: Penguin Press, 2023.

Tussman, Joseph. "The Collegiate Rite of Passage." In *Experiment and Innovation: New Directions in Education at the University of California*, July 1968.

U.S. Department of Defense. *Military Specification Cookie Mix Dry*, MIL-C-43205G.

U.S. Department of Energy. *The Manhattan Project: Making the Atomic Bomb*. January 1999.

U.S. Department of the Interior. National Register of Historic Places Inventory. *Historic Hutterite Colonies Thematic Resources*. 1979.

U.S. Department of the Treasury. "Treasury Identifies Eight Chinese Tech Firms as Part of the Chinese Military-Industrial Complex." December 16, 2021. home.treasury.gov/news/press-releases/jy0538.

U.S. National Institute of Standards and Technology. "Face Recognition Technology Evaluation (FRTE) 1:1 Verification." 2024.

Vance, Ashlee. *Elon Musk: Tesla, SpaceX, and the Quest for a Fantastic Future*. New York: HarperCollins, 2015.

Veblen, Thorstein. *The Theory of the Leisure Class*, edited by Martha Banta. New York: Oxford University Press, 2007.

Vinograd, Cassandra, and Isabel Kershner. "Israel's Attackers Took About 240 Hostages." *New York Times*, November 20, 2023.

Voltaire. *Zadig; or, The Book of Fate*. London, 1749.

Wakefield, Dan. "William F. Buckley Jr.: Portrait of a Complainer." *Esquire*, January 1, 1961.

Wallace, George C. "Inaugural Address." Montgomery, Ala., January 14, 1963, Alabama Department of Archives and History.

Wallace, Robin. "Why Beethoven's Loss of Hearing Added Dimensions to His Music." *Zócalo Public Square*, July 28, 2019.

Watlington, Emily. "'Monet/Mitchell' Shows How the Impressionist's Blindness Charted a Path for Abstraction." *Art in America*, May 12, 2023.

Weisgerber, Marcus. "F-35 Production Set to Quadruple as Massive Factory Retools." *Defense One*, May 6, 2016.

Wendling, Mike. "Xi Jinping: Chinese Leader's Surprising Ties to Rural Iowa." BBC, November 15, 2023.

Whang, Oliver. "How to Tell if Your A.I. Is Conscious." *New York Times*, September 18, 2023.

White, Debbie. "Drones Branch Out To Swarming Through Forests." *Times* (London), May 5, 2022.

White, Richard D., Jr. "Executive Reorganization, Theodore Roosevelt, and the Keep Commission." *Administrative Theory and Praxis* 24, no. 3 (2002): 507–18.

Whyte, Kenneth. *Hoover: An Extraordinary Life in Extraordinary Times*. New York: Alfred A. Knopf, 2017.

Wigner, Eugene. "The Unreasonable Effectiveness of Mathematics in the Natural Sciences." *Communications in Pure and Applied Mathematics* 13, no. 1 (February 1960): 1–14.

Winston, Ali. "Palantir Has Secretly Been Using New Orleans to Test Its Predictive Policing Technology." *Verge*, February 27, 2018.

Wong, Julia Carrie. "'We Won't Be War Profiteers': Microsoft Workers Protest $48M Army Contract." *Guardian*, February 22, 2019.

Wortman, Marc. *Admiral Hyman Rickover: Engineer of Power*. New Haven, Conn.: Yale University Press, 2022.

Yglesias, Matthew. "Pay Congress More." *Vox*, May 10, 2019.

Yudkowsky, Eliezer. "Pausing AI Development Isn't Enough. We Need to Shut It All Down." *Time*, March 29, 2023.

Zachary, G. Pascal. *Endless Frontier: Vannevar Bush, Engineer of the American Century*. New York: Free Press, 1997.

Zelinsky, Nathaniel. "Challenging the Unchallengeable (Sort Of)." *Yale Alumni Magazine*, January/February 2015.

Zhang, Taisu, Graham Webster, and Orville Schell. "What Xi Jinping's Seattle Speech Might Mean for the U.S." *Foreign Policy*, September 23, 2015.

Zhou, Xin, et al. "Swarm of Micro Flying Robots in the Wild." *Science Robotics* 7, no. 66 (2022).

Art Credits

All images have been either reprinted or redrawn
based on data from the following sources:

Figure 1: Sébastien Bubeck et al., "Sparks of Artificial General Intelligence: Early Experiments with GPT-4," arXiv, March 22, 2023, 7.

Figure 2: Steven Pinker, *Enlightenment Now: The Case for Reason, Science, Humanism, and Progress* (New York: Penguin Books, 2018), 159.

Figure 3: World Bank Group, "Military Expenditures (% of GDP): United States, European Union, 1960–2022."

Figure 4: Robert J. Gordon, *The Rise and Fall of American Growth* (Princeton, N.J.: Princeton University Press, 2016), 547.

Figure 5: Aden Barton, "How Harvard Careerism Killed the Classroom," *Harvard Crimson*, April 21, 2023 (citing Claudia Goldin et al., "Harvard and Beyond Project," Harvard University, 2023).

Figure 6: Angus Maddison, *Contours of the World Economy, 1–2030 AD: Essays in Macro-Economic History* (Oxford: Oxford University Press, 2007), 70.

Figure 7: Samuel P. Huntington, "The Clash of Civilizations?," *Foreign Affairs* 72, no. 3 (Summer 1993): 30 (citing William Wallace, *The Transformation of Western Europe* (London: Pinter, 1990)).

Figure 8: Niall Ferguson, *Civilization: The West and the Rest* (New York: Penguin Books, 2011), 6.

Figure 9: Martin Lindauer, "House-Hunting by Honey Bee Swarms," trans. P. Kirk Visscher, Karin Behrens, and Susanne Kuehnholz, *Journal of Comparative Physiology* 37 (1955): 274.

Figure 10: Solomon E. Asch, "Opinions and Social Pressure," *Scientific American* 193, no. 5 (November 1955): 32. Reproduced with permission. Copyright © 1955 Scientific American, Inc. All rights reserved.

Figure 11: Drew Desilver, "New Congress Will Have A Few More Veterans, But Their Share of Lawmakers Is Still Near A Record Low," *Pew Research Center*, December 7, 2022.

Figure 12: Philip E. Tetlock, *Expert Political Judgment: How Good Is It? How Can We Know?* (Princeton, N.J.: Princeton University Press, 2005), 77.

Figure 13: Tom Giratikanon et al., "Up Close on Baseball's Borders," *New York Times*, April 24, 2014. From *The New York Times*. © 2014 The New York Times Company. All rights reserved. Used under license.

Figure 14: Herbert James Draper, *Ulysses and the Sirens*, 1909, oil on canvas, 177 × 213.5 cm, Ferens Art Gallery, Kingston Upon Hull, England.

Figure 15: Chris Zook, "Founder-Led Companies Outperform the Rest—Here's Why," *Harvard Business Review*, March 24, 2016.

Index

accommodation, 156–57
Adams, John, 6
Adekoya, Remi, 53
aesthetics, 205–8, 209
Afghanistan, war in, 139–45, 152–54
Agnew, Spiro, 66
aircraft, military, 44–45, 76–77, 141–42
Alexander the Great, 139
Alphabet, 77
Altman, Sam, 47
altruism movement, 213
Amar, Akhil Reed, 212
Amazon, 77, 111, 175
American Civil Liberties Union (ACLU), 57–58, 175
American Culture, 194
American Historical Association, 88
amorality, 17
Anderson, Benedict, 191, 198
anonymity, 67
Appiah, Kwame Anthony, 85, 89
Apple, 77, 100–102
Applebaum, Anne, 51–52
Archilochus, 160
aristocracies, 78–79
Arouet, François-Marie (Voltaire), 173
art criticism, 205–6

Art Students League of New York, 156
artificial intelligence
  backlash against, 24–25
  capabilities of, 19–21
  deterrence and, 28
  doubts regarding, 24
  ethics and, 18
  large language models and, 11–12, 18–19, 22–23
  law enforcement and, 173, 175
  military applications and, 31–32, 45–48
  predictions regarding, 25–26
  U.S. Defense Department and, 12
  weapons systems and, 26
Asch, Solomon E., 13, 13–14, 130–32, 138
Asylums (Goffman), 65
atomic/nuclear weapons, 16–17, 35–36, 37, 39
authoritarian regimes, 51–52

BAE Systems, 173
Baldwin, James, 173
Baltzell, E. Digby, 78–79
baseball, 193*fig*
Basquiat, Jean-Michel, 106–7
battle-related deaths, 40*fig*

Beard, Mary, 206
bees, study of swarm of, 115–19
Beethoven, Ludwig van, 137–38
Before Sunset, 105
Bellah, Robert N., 200–201
Benedict, Ruth, 212
Benedict, Saint, 188–89
Benton, Thomas Hart, 156
Berlin, Isaiah, 159–60, 164
Berman, Morris, 81
Bermingham, Edward J., 43
Bethe, Hans, 7
Better Angels of Our Nature, The (Pinker), 40–41
biblical archetypes, 201
"Big Idea Famine" (Negroponte), 48
Binet, Alfred, 116
Bing, 22–23
Black Panthers, 65–66
Blackstone, William, 173
Bloom, Allan, 31, 66
Booker, Cory, 176
Borrell, Josep, 42
Brand, Stewart, 98
Brennan, Timothy, 91
Brewster, Kingman, Jr., 58, 59, 65–66
Bridgman, Percy Williams, 17
Brill, Steven, 151–52
Brooks, David, 177
Bruckner, Pascal, 64
Bubeck, Sébastien, 19
Bubis, Ignatz, 203–4
Buckley, William F., Jr., 34
"buffalo killers," 140
Burke, Kenneth, 188
Bush, Vannevar, 4–5, 7–8, 37, 95, 99
Butz, James, 139

Caddell, Patrick, 145
Cameron, David, 42
career paths, changes in, 75–76, 75*fig*
Carlyle, Thomas, 197
Carlyle Group, 179
Carter, Jimmy, 185
Carter, Stephen L., 72
caste structures, 79
Castells Oliván, Manuel, 70
Catiline, 215
Central Intelligence Agency, 4
ChatGPT, 23, 47
Checkers speech (Nixon), 62
Cheyette, Fredric L., 83, 85, 86–87
China, censorship and, 67
Chomsky, Noam, 24
Churchill, Winston, 89
civil rights movement, 98
Civilisation, 205–6
Clark, Kenneth, 205–6
"Clash of Civilizations, The" (Huntington), 85
classical liberalism, 67–68
Clinton, Bill, 149–50, 152
Closing of the American Mind, The (Bloom), 66
CloudWalk Technology, 30
collective decision making, 115–21
collective identity, 13
Collin, Frank, 57
Coming of Age in Samoa (Mead), 213n
computers, personal, 4, 98–99, 101–2
conformity, 130–32, 131*fig*, 138, 158–59
Congress, 144–45, 144*fig*, 180–82, 185–86
consciousness, study of, 23–24
constructive disobedience, 136
consumer culture, impact of, 47

consumer market, 9–10, 13, 96, 103–6, 110, 172–73
corporate culture, 124–26, 127–28
corruption, 185–88
Cortés, Hernán, 70n
counterculture movement, 98, 100
creation for creation's sake, 71
creativity, 19–20, 158
crime, 173–77
crowds, wisdom of, 171–72
Cultural Revolution, 51
Culture of Disbelief, The (Carter), 72
Curie, Marie, 8

"dance language," 117
"decoupling," 95
Deep Blue, 20–21
DeepMind, 21
Defense Advanced Research Projects Agency, 74–75
defense spending, 41–43, 42*fig*
Delpy, Julie, 105
Denby, David, 206
Descartes, René, 24
deterrence, doctrine of, 41
Dewey, John, 161
disability, adaptation and, 137–38
disclosure, political candidates and, 62–63
disruption, 106
dot-com bubble, 108–9
Dowd, Maureen, 60
draft, military, 144, 145
Draper, Herbert James, 208*fig*
drone swarms, 30
Drucker, Peter F., 127
Dunbar, Robin, 190–91
Dunbar's number, 190–91
dynamite, 37–38

Eck Swarm, 115–19, 118*fig*
Economic Club of Washington, D.C., 179
egalitarianism, pursuit of, 215–16
Einstein, Albert, 8, 35
Eisenhower, Dwight D., 7, 43, 71
Emerson, Ralph Waldo, 66, 159
End of History, The (Fukuyama), 31
End Times (Turchin), 107
engineering culture, 160–61, 166
Enlightenment, 214
Entrepreneurial State, The (Mazzucato), 74–75
estimates, groups and, 171
ethical universalism, 213
ethnographic present, 213n
eToys, 103–5, 108–9
Europe, GDP spent on defense by, 41–43, 42*fig*
Expert Political Judgment (Tetlock), 162

Facebook, 71, 111
facial recognition, 29–30, 176
Fahlenbrach, Rüdiger, 209–10
Fairchild Camera and Instrument Corporation, 4
Fan Hui, 21
FarmVille, 154
Faust (Goethe), 64
Federal Acquisition Streamlining Act (1994), 151, 152–53
Federal Bureau of Investigation (FBI), 101, 173
Federal Reserve, 179–80
Federal Trade Commission, 25
Felsenstein, Lee, 98, 100
Ferguson, Niall, 93
First Amendment, 57–59

Five Whys, 164–66
Florentius, 188–89
Ford Aerospace, 4
founder culture, 178
founder-led companies, 209–11, 210*fig*
"foxiness," 163–64
Franklin, Benjamin, 5
free speech, protection of, 57–60
Freire, Paulo, 54
Freud, Lucian, 166–67
Freud, Sigmund, 72, 166–67
Friedan, Betty, 58–59
Friedman, Richard Alan, 157
Frisch, Karl von, 117
Fukuyama, Francis, 31, 73

Gaddis, John Lewis, 40
gait recognition systems, 173
Galton, Francis, 171
Gay, Claudine, 64
Gayford, Martin, 167
Gaza, 60, 67
GDP, 78*fig*, 197
General Dynamics Corporation, 185
German Publishers and Booksellers Association, 201–2
German reunification, 43
Girard, René, 157–58
Glenn, John, 151
global economic production, 93–94, 94*fig*
Go, 21
Goethe, 64
Goffman, Erving, 65
Goh Keng Swee, 196
Goh Report, 196
Goldberg, Jeffrey, 42
Good, Irving John, 26
Google, 22, 33, 64, 74, 111

Gopinathan, Saravanan, 196
Gordon, Robert J., 48–49
Gore, Al, 148, 150
Gotham, 175
government procurement process, 146–52
GPT-4, 19, 25
Graeber, David, 47, 110
Grass, Günther, 43
Great Man theory, 197–98, 206
Greene, Diane, 33
Greenspan, Alan, 110
grievance, 157
Groupon, 154, 172
Gruber, Howard, 137
Grumman Corporation, 141–42
Gulf War, 145–47, 150
Gutmann, Amy, 69–70

Habermas, Jürgen, 8
Hackers: Heroes of the Computer Revolution (Levy), 99
Harris, Kamala, 176
Harvard University, 75–76, 75*fig*
Harvey, Paul, 63n
Hawke, Ethan, 105
Hay, Denys, 92
"heckler's veto," 59
Hedgehog and the Fox, The (Berlin), 159–60
Heller, Ágnes, 80
Hemingway, Ernest, 52
Henderson, Rob, 177
Herman, Arthur L., 142
Herschbach, Dudley, 5
Hersey, John, 38
Herzog, Roman, 203
Hindenburg, Paul von, 39
Hitler, Adolf, 4, 39, 202

INDEX

Hofstadter, Douglas, 24
Homebrew Computer Club, 98
Homer, 208n
Hoover, Herbert, 160
Hoover, J. Edgar, 173
Horn, Marian Blank, 153
Huntington, Samuel, 85–86
Huntington-Wallace Line, 85, 86*fig*
Hutterites, 190

IBM, 20–21, 99–100, 101, 176
Impro (Johnstone), 123
improvised explosive devices (IEDs), 139–42
inclusivity, 192
indigenous peoples, 53
innovation deserts, 14
Instagram, 50
intellectual courage, 73
iPhone, 50, 101
"irrational exuberance," 110–11
Isaacson, Walter, 100
Israel, 60, 67

jackdaws, 124
Japan, pacifism of, 43–44
Jefferson, Thomas, 5
Jégo, Yves, 192
Jobs, Steve, 100–101
Johnstone, Keith, 122–24
judgment, suspension of, 167

Kasparov, Garry, 21
Keeley, Lawrence H., 53
Kennedy, Robert F., 87
Kerouac, Jack, 157
Khan, Lina, 25
Kimball, Roger, 201
King, Charles, 213n

King, Martin Luther, Jr., 87
King Solomon's Ring (Lorenz), 124
Kintner, Edwin E., 183
Kissinger, Henry, 52, 76, 197
Kris, Ernst, 158n
Kristol, Irving, 215
Ku Klux Klan, 58

LaMDA, 22
language, 191, 195–97
large language models, 11–12, 18–20, 21–25, 26–27
law enforcement, 173–77
Le Pen, Jean-Marie, 193
Leclerc, Georges-Louis, 5
Lee, Richard C., 58
Lee Kuan Yew, 182, 194–95, 196–97, 215
Lehman, John, 186
Lemoine, Blake, 22
Lenk, Toby, 103–5, 106, 109
Lévi-Strauss, Claude, 89n
Levy, Steven, 99–100
Liberalism and the Limits of Justice (Sandel), 68
Licklider, Joseph, 7
lifestyle technology, 107
Lindauer, Martin, 115–19
Link, Perry, 66–67
Linklater, Richard, 105
Lockheed Martin, 44–45, 76–77, 141, 142
Lockheed Missile & Space, 4
"long peace," 40–41, 40*fig*
Lord of the Rings, The (Tolkien), 199
Lorenz, Konrad, 124
Ludendorff, Erich, 39
Lundberg, Ferdinand, 78n
"luxury beliefs," 177

machine learning systems, 32–33
Macintosh computers, 101–2
MacIntyre, Alasdair, 201
Macron, Emmanuel, 191–92
Madison, James, 5, 181
Magill, Elizabeth, 60, 64
"Man-Computer Symbiosis" (Licklider), 7
Manhattan Project, 16, 46
"market triumphalism," 172
Mattis, James, 143
Mazzucato, Mariana, 74–75
McCain, John, 153
McCarthy, Joseph R., 81
McNeill, William, 84, 93
Mead, Margaret, 213n
meetings, pervasiveness of, 126
Mencken, H. L., 39
Merkel, Angela, 6n
Meta, 71, 77
Meyer, Edith Patterson, 38
Microsoft, 22–23, 33, 77
Middle Americans, as person of the year, 97
Milgram, Stanley, 13–14, 132–36
military applications, 31–32, 33–34, 63
military-industrial complex, 71
Milley, Mark, 44–45
Mills, D. Quinn, 109
mimicry, 158
Mishra, Pankaj, 90–91, 93
Mitchell, Joan, 137
Mölling, Christian, 43
Monet, Claude, 137
monkeys, conflicts among, 157–58
moral allegiance, 69–70
moral dualism, 53
moral obtuseness, 214

Morin, Chloé, 53
Motorola, 146, 150
Mulaney, John, 156
Munich Zoological Institute, 116–17
Murray, Pauli, 58–59
Murray, Thomas E., 183
Murrow, Edward R., 81
Musk, Elon, 49

Nadella, Satya, 33
NASA, 7
national cultures/identities, 191–201, 204, 216–18
National Front party, 193
National Institute of Standards and Technology, 29–30
National Organization for Women, 59
National Physical Laboratory, 173
Nautilus, USS, 184
Nazi Party, 57–58, 131–32, 161, 202
Negroponte, Nicholas, 48
Neier, Aryeh, 57–58, 59
New Orleans Police Department, 173–74
1984 (Orwell), 79–80
"1984" advertising campaign, 101
Nixon, Richard, 62
Nobel, Alfred, 37–38
Noonan, Peggy, 23, 25, 177, 206
North Atlantic Treaty Organization (NATO), 43
Northrop Grumman, 141
nuclear/atomic weapons, 16–17, 35–36, 37, 39
nuclear-powered submarines, 183–84
Nussbaum, Martha, 198
Nvidia, 77
Nye, Joseph S., Jr., 32n

Obama, Barack, 41–42
obedience experiment, 132–36
observation, 166–67
"Obsessive Actions and Religious Practices" (Freud), 72
Odyssey (Homer), 208n, 209
Ohno, Taiichi, 164
Old Man and the Sea, The (Hemingway), 52
On the Road (Kerouac), 157
OpenAI, 47
Oppenheimer, J. Robert, 7, 16–17, 37, 99
optionality, pursuit of, 69–70
Orientalism (Said), 90–93
Orton, Brad, 147
Orwell, George, 79–80
Osnos, Evan, 51
ownership culture/societies, 178, 211

pacifism, 43–44, 53, 54
Palantir, 13–14, 26, 63, 122–23, 125, 137, 142–43, 152–55, 165–66, 173–74, 199
Panthéon, 197
Parisi, Giorgio, 120–21
Parsons, Talcott, 107–8
Patterns of Culture (Benedict), 212
Paul, Weiss, Rifkind, Wharton & Garrison, 58
Pedagogy of the Oppressed (Freire), 54
Permanence and Change (Burke), 188
Philco, 125
Pichai, Sundar, 33
Pinchot, Gifford, 148
Pinker, Steven, 40–41
Places of Mind (Brennan), 91
Planck, Max, 72
Plato, 188n

political candidates, 61, 62–63
Pollock, Jackson, 156
Pope Gregory, 188–89
postcolonial studies, 91
postmodernism, 73, 81
Powell, Jerome, 179–80
"Pragmatic America" (Dewey), 161
Pratt & Whitney, 77
predictions, study of, 162–64, 163*fig*
Priceline, 109
productivity, 48–49
productization, 72
Project Maven, 33
Project Y, 16
Protestant Establishment, The (Baltzell), 78
Proxmire, William, 186
public office, as overmoralized, 61–62
public sector compensation, 179–82
Putin, Vladimir, 43

radios, Gulf War and, 145–47, 150
Ram Trucks, 63n
Rangel, Charles, 144–45
Rapid Equipping Force, 142
Rawls, John, 216n
Raytheon, 153
religion, 72–73, 159, 200–201
Renan, Ernest, 199
Republic, The (Plato), 188n
Rickover, Hyman G., 184–86, 187
Roosevelt, Franklin, 4–5, 35
Roosevelt, Theodore, 148
Rosenbaum, David E., 150
Roth, William, 147
Rubenstein, David, 179

Sabato, Larry, 61
Said, Edward, 90–93

Sallust, 215
Sandel, Michael, 67–68, 172
Sawyer, Diane, 184
scaling, 71
scapegoat mechanism, 188
Schelling, Thomas, 32
science, faith in, 73
scientific revolution, 214
Scott, Ridley, 101
Scruton, Roger, 213
Sculley, John, 101
Seeley, Thomas D., 115
Seinfeld, Jerry, 122
"Self-Reliance" (Emerson), 159
Sennett, Richard, 198
shared ownership models, 211–12
Shatz, Adam, 90
Sherick, Joseph, 149
Silicon Valley
   aesthetics and, 207–8
   altruism movement and, 213
   as base of technology sector, 77
   conformity and, 138
   consumer market and, 172–73
   culture of, 128–29, 136
   idealism and, 213–14
   individualism and, 99
   military applications and, 46–48, 74–75
   modern incarnation of, 9–10
   national culture and, 216–18
   ownership culture and, 211–12
   rise of, 3–4
   salaries in, 187
Silver, Nate, 95
Simon, Herbert A., 25–26
Singapore, 194–97
Singer, Peter, 213

16th Street Baptist Church bombing, 58
Smith, Brad, 33
Smith, James K. A., 200
"So God Made a Farmer" (Harvey), 63n
social deafness, 137
social media, growth of, 50
Social Network, The, 71
social relationships, group size and, 190–91
software industry, 3–4, 9–10, 11, 32–34, 160, 173–76
Somerville, Mary, 6
Soviet Union, proscriptions of, 66–67
space program, 7
SpaceX, 49
sports, 192, 193, 193*fig*
Sputnik, 7
Starkie, Thomas, 173
starlings, 120–21
startups, 120, 122–29, 136
Stasi, 80
status, 123–25
Strauss, Leo, 214, 215
Summers, Lawrence, 65
Swensen, David, 211
Szilard, Leo, 35

Talmud, 29
Tan Dan Feng, 196
Tesla, 49, 77
Tet Offensive, 87–88
Tetlock, Philip E., 162–64, 163*fig*
Thatcher, Margaret, 6n
Thiel, Peter, 50
Thoreau, Henry David, 52
"Three Wooden Crosses," 200
Time magazine, 97

INDEX

Tolkien, J. R. R., 199
"total institutions," 65
total war, 39
Toyota Motor Corporation, 164
transparency, 61
Travis, Randy, 200
trigger warnings, 157
Turchin, Peter, 107
Tussman, Joseph, 88
Twain, Mark, 52
Twilight of American Culture, The (Berman), 81
twin studies, 29–30

Ukraine, Russia's invasion of, 43
Ulysses and the Sirens (Draper), 208*fig*
United States
  GDP spent on defense by, 41–43, 42*fig*
  national culture and, 192–94, 216
United Technologies, 4
university presidents, Congress and, 60, 64–65
U.S. Air Force, 145–47
U.S. Atomic Energy Commission, 183
U.S. Defense Advanced Research Projects Agency, 7
U.S. Defense Department, 12, 33, 45
U.S. government, rise of Silicon Valley and, 3–4
U.S. Navy, 183–86

Vatican library, 6
Veblen, Thorstein, 206–7

Vietnam War, 87–88, 98
virtues, conceptions of, 215
Voltaire, 173

Wallace, George, 58–59
Wallace, William, 85
Walser, Martin, 201–4
Walt Disney Company, 103
War Before Civilization (Keeley), 53
Weeping Willow (Monet), 137
West
  challenging of, 95–96, 98
  concept of, 89–90
Western civilization courses, 83–85, 87–88, 93
Westinghouse, 4
"What Is a Nation?" (Renan), 199
Whitman, Walt, 52
Whole Earth Catalog, 98
Wigner, Eugene, 162
WilmerHale, 64
World War II, 4–5, 7–8, 38–39, 87, 141–42, 161, 202–3

Xi Jinping, 51–53
Xi Mingze, 52

Yale Political Union, 58, 59
Yglesias, Matthew, 181
YouTube, 50
Yudkowsky, Eliezer, 25

Zuckerberg, Mark, 71
Zynga, 154, 172

penguin.co.uk/vintage